SPRINGER LABORATORY

Springer

Berlin
Heidelberg
New York
Hong Kong
London
Milano
Paris
Tokyo

Koichi Hatada · Tatsuki Kitayama

NMR Spectroscopy of Polymers

With 110 Figures and 43 Tables

Springer

Prof. Dr. Koichi Hatada

Prof. Dr. Tatsuki Kitayama

Graduate School of Engineering Science
Department of Chemistry
Osaka University
Machikaneyama 1–3
Toyonaka
560-8531 Osaka, Japan

hatada@chem.es.osaka-u.ac.jp
kitayama@chem.es.osaka-u.ac.jp

ISBN 978-3-642-07293-2

Library of Congress Cataloging-in-Publication-Data applied for

A catalog record for this book is available from the Library of Congress.
Bibliographic information published by Die Deutsche Bibliothek
Die Deutsche Bibliothek lists this publication in the Deutsche Nationalbibliographie; detailed
bibliographic data is available in the internet at <http://dnb.ddb.de>.

Springer-Verlag is a part of Springer Science+Business Media

springeronline.com

© Springer-Verlag Berlin Heidelberg 2010
Printed in Germany

Coverdesign: Künkel & Lopka, Heidelberg

0/2/3020/ ra – 5 4 3 2 1 0   Printed on acid-free paper

Laboratory Manual Series in Polymer Science

Editors

Prof. Howard G. Barth
DuPont Company
P.O. Box 80228
Wilmington, DE 19880-0228
USA
howard.g.barth@usa.dupont.com

Priv.-Doz. Dr. Harald Pasch
Deutsches Kunststoff-Institut
Abt. Analytik
Schloßgartenstr. 6
64289 Darmstadt
Germany
hpasch@dki.tu-darmstadt.de

Editorial Board

PD Dr. Ingo Alig
Deutsches Kunststoff-Institut
Abt. Physik
Schlossgartenstr. 6
64289 Darmstadt
Germany
ialig@dki.tu-darmstadt.de

Prof. Josef Janca
Université de La Rochelle
Pole Sciences et Technologie
Avenue Michel Crépeau
17042 La rochelle Cedex 01
France
jjanca@univ.lr.fr

Prof. W.-M. Kulicke
Inst. f. Technische u. Makromol. Chemie
Universität Hamburg
Bundesstr. 45
20146 Hamburg
Germany
kulicke@chemie.uni-hamburg.de

Prof. H. W. Siesler
Physikalische Chemie
Universität Essen
Schützenbahn 70
45117 Essen
Germany
hw.siesler@uni-essen.de

Springer Laboratory Manuals in Polymer Science

Pasch, H.; Trathnigg, B.: HPLC of Polymers
ISBN: 3-540-61689-9 (hardcover)
ISBN: 3-540-65551-4 (softcover)

Mori, S.; Barth, H.G.: Size Exclusion Chromatography
ISBN: 3-540-65635-9

Pasch, H.; Schrepp, W.: MALDI-TOF Mass Spectrometry of Synthetic Polymers
ISBN: 3-540-44259-6

Kulicke, W.-M.; Clasen, C.: Viscosimetry of Polymers and Polyelectrolytes
ISBN: 3-540-40760-X

Hatada, K.; Kitayama, T.: NMR Spectroscopy of Polymers
ISBN: 3-540-40220-9

Preface

Nuclear magnetic resonance (NMR) spectroscopy is one of the most widely and frequently used methods for structure analysis in chemical research. So many books and monographs on NMR spectroscopy have been published aiming at the fundamental understanding as well as the practical applications especially to organic chemistry and biochemistry. Polymer chemistry has also been benefited from NMR spectroscopy but in a somewhat different and specialized manner, such as tacticity determination, monomer sequence analysis of copolymers, analysis of end groups and irregular linkages, and chain dynamics of polymers and so on. Therefore another NMR book which is oriented to practical use of the spectroscopy in polymer chemistry should be added to the list of NMR text books.

This book places emphasis on the practical use of NMR spectroscopy in polymer chemistry rather than on the theoretical treatments. In the first chapter, after the description of fundamental aspects of NMR spectroscopy, experimental problems such as preparation of sample solutions, selection of the solvent, internal standards and tubing, and contaminants in the sample solution are discussed. The second chapter is devoted to discussion on the accuracy and precision of NMR measurements, since a much higher degree of accuracy and precision is required in the analysis of polymer structures, such as tacticity, copolymer composition, and chain-end structures. This chapter also includes the explanation of the coaxial tubing method, which is very useful for quantitative analysis and determination of volume magnetic susceptibility by NMR.

Chapters 3–5 describe structural analysis of polymers, dealing with the stereo-chemistry of polymer chains (Chap. 3), chemical composition and comonomer sequence distribution in copolymers including diene polymers (Chap. 4), and end groups and irregular linkages (Chap. 5). The analysis of polymerization reactions by NMR and the relationship between chemical shift and reactivity for vinyl monomers are also described in Chap. 5. This information is quite useful for the understanding of polymerization reaction, and is discussed in this connection.

Two-dimensional NMR spectroscopy is introduced in Chap. 6 with examples of the application to polymer and oligomer analysis, including conformational analysis of methyl methacrylate oligomers.

NMR spectroscopy is also a powerful tool for the investigation of polymer chain dynamics in solution by the aid of NMR relaxation parameters, including T_1 and nuclear Overhauser enhancement (NOE). The problem is discussed in Chap. 7, in which the precision and effect of experimental conditions in the determinations of relaxation times and NOE are discussed.

Combined use of spectroscopy and chromatography is one of the promising trends in analytical chemistry and thus we add one chapter (Chap. 8) describing on-line coupled size-exclusion chromatography (SEC)/NMR spectroscopy in which an NMR spectrometer is set in the SEC system as a detector. The system allows fast and facile determinations of the molecular weight dependence of polymer characteristics, such as tacticity and copolymer composition as well as the molecular weight itself.

The authors strongly hope that the basics in NMR measurements described in this book will be helpful and useful for many NMR users as well as newcomers to the field of NMR. An well-ordered index and a list of abbreviations are appended for the reader's convenience.

The contents of this book largely come from the authors' experiences in research work in polymer chemistry carried out at the Faculty of Engineering Science, Osaka University, where one of the authors (K.H.) first encountered a 100 MHz NMR spectrometer in 1965. The NMR research in the Faculty of Engineering Science has been cultivated by collaboration with the faculty members who have been actively involved in obtaining high-quality NMR data from time to time: Mr. Yoshio Terawaki (since 1965), Mr. Hiroshi Okuda (since 1968), and Dr. Koichi Ute (since 1985). Thus our most sincere thanks should be extended first to these people. We are particularly grateful to Mr. Terawaki, who has devoted himself to collecting NMR data for this book and also in assisting in the preparation of the manuscript.

The authors are also indebted to people who participated in round-robin tests on polymer samples which were organized by the Research Group on NMR, the Society of Polymer Science, Japan, the outcomes of which constitute important parts of several chapters in this book. K.H. is particularly grateful to Professor Riichiro Chûjô and Professor Yasuyuki Tanaka, who were the cofounders with K.H. of the research group, for their continuous encouragement and friendship.

During the preparation of the manuscript, the members of the Hatada and Kitayama laboratories were very helpful, and, in particular, Mrs. Fumiko Yano and Mr. Takafumi Nishiura, who typed and prepared the manuscript, Dr. Takehiro Kawauchi, who prepared many of the figures in this book, Mr. Ken-ichi Katsukawa, and Dr. Hidetaka Ohnuma are greatly acknowledged.

The authors are grateful to Professor Isao Ando, Tokyo Institute of Technology, who reviewed the entire manuscript and gave valuable advice on many problems. They also thank the editors of Springer-Verlag for their friendly cooperation in seeing this work through the throes of final production.

Osaka, November 2003   K. Hatada
                                                                                                T. Kitayama

List of Symbols and Abbreviations

ADC	analogue-to-digital converter
AIBN	azobis(isobutyronitrile)
AN	acrylonitrile
B_0	static magnetic field
B_1	radio frequency field
BF	broadening factor
BPO	benzoyl peroxide
BuMA	n-butyl methacrylate
COM	complete decoupling condition
COSY	correlation spectroscopy
CW	continuous wave
DEPT	distortionless enhancement of polarization transfer
DMSO	dimethyl sulfoxide
DP	degree of polymerization
DQF-COSY	double-quantum filtered correlation spectroscopy
DSS	4,4-dimethyl-4-silapentane sulfonate
EMA	ethyl methacrylate
EPDM	ethylene–propylene–(2-ethylidene-5-norbornene) terpolymer
Eu(fod)$_3$	tris[1,1,1,2,2,3,3-heptafluoro-7,7-dimethyloctanedionato(4,6)]-europium(III)
Eu(tfmc)$_3$	tris[(3-trifluoromethylhydroxymethylene)-(+)-camphorato]-europium(III)
FID	free induction decay
FT	Fourier transform
GPC	gel permeation chromatography, size exclusion chromatography (SEC)
h	Planck constant
HETCOR	heteronuclear chemical shift correlation spectroscopy
HFA	hexafluoroacetone
HMBC	heteronuclear multiple-bond correlation spectroscopy
HMDS	hexamethyldisiloxane
HMDS'	hexamethyldisilane
HMQC	heteronuclear multiquantum correlation spectroscopy
HPLC	high-performance liquid chromatography
HSC	heteronuclear shift correlation spectroscopy
HSQC	heteronuclear single-quantum correlation spectroscopy
I	nuclear spin
INADEQUATE	incredible natural abundance double quantum transfer experiment
J	spin–spin coupling constant
k	Boltzmann constant

M	magnetization vector
m	magnetic quantum number
m diad	meso diad
MMA	methyl methacrylate
Mn	number-average molecular weight
MWD	molecular weight distribution
N	number of data points
NMR	nuclear magnetic resonance
NNE	gated decoupling mode with suppression of NOE
NOE	nuclear Overhauser enhancement
NOESY	2D nuclear Overhauser enhancement spectroscopy
$N_\alpha(N_\beta)$	number of α spins (β spins) (α: lower-energy state, β: upper-energy state)
OMTS	octamethylcyclotetrasiloxane
PEMA	poly(ethyl methacrylate)
PMMA	poly(methyl methacrylate)
PMVE	poly(methyl vinyl ether)
ppm	parts per million
r	racemo diad
RF	radio frequency
r_1, r_2	monomer reactivity ratios
RI	refractive index
S/N	signal-to-noise ratio
SEC	size-exclusion chromatography, gel permeation chromatography (GPC)
SPSJ	Society of Polymer Science, Japan
T_1	spin–lattice relaxation time
T_2	spin–spin relaxation time
THF	tetrahydrofuran
TMS	tetramethylsilane
TSP-d_4	sodium [2,2,3,3-d_4]3-trimethylsilylpropanoate
VPO	vapor pressure osmometry
WET	water suppression enhanced through T_1 effects
wt/vol%	weight of sample (g)/volume of solution (ml) %
γ	gyromagnetic ratio
δ	chemical shift (ppm)
μ	nuclear magnetic moment
υ	frequency
σ	shielding constant
σ	standard deviation
τ_c	correlation time
χ	volume magnetic susceptibility
ω	angular frequency

Table of Contents

1 Introduction to NMR Spectroscopy

1.1 Basic Principles of NMR

This chapter describes basic principles of nuclear magnetic resonance (NMR) and the relating terms which represent the minimum knowledge required to read this book. For further details, readers are advised to consult NMR text books [1-5] and an encyclopedia [6]. Periodical reviews on NMR spectroscopy have also been published [7,8]. The chapter also describes some details of NMR experiments, including tips in sample preparations with particular emphasis on polymer samples, and is concluded by a section outlining NMR measurements.

1.1.1 Nuclear Magnetic Resonance

The fundamental property of the atomic nucleus involved in NMR is the nuclear spin, I, which has values of 0, 1/2, 1, 3/2, etc., in units of $h/2\pi$. The value of the spin depends on the mass number and the atomic number of the nucleus. Nuclei with $I=0$ (^{12}C, ^{16}O, ^{32}S,...) have no spin angular momentum and thus no magnetic moment, and NMR experiments cannot be conducted. Examples of nuclei that have $I=1/2$ are the normal (natural abundance $\cong 100\%$) nuclei of ^1H, ^{19}F, and ^{31}P; other less abundant nuclei, such as those of ^{13}C, ^{15}N, and ^{29}Si, also have $I=1/2$.

The nuclear magnetic moment, μ, is given as follows:

$$\mu = \gamma I\, h/2\pi, \tag{1.1}$$

where h is Planck's constant and γ is the gyromagnetic ratio and is a constant for each particular nucleus. The nuclear magnetic moments, when placed in a uniform magnetic field, B_0, orient themselves with only certain allowed orientations as quantum mechanics tells us; a nucleus of spin I has $(2I+1)$ possible orientations. The nuclei commonly observed in NMR spectroscopy of organic compounds (e.g., ^1H and ^{13}C) have spin $I=1/2$ and thus two magnetic states, characterized by a set of magnetic quantum numbers $m=1/2$ and $-1/2$, (in general $m=I, I-1, I-2,..., -I$). The magnetic moment μ is aligned either parallel to the magnetic field, B_0, ($m=1/2$) or antiparallel to B_0 ($m=-1/2$). The energy of the spin states is proportional to μ, and B_0 and is written as Eq. (1.2):

$$E = -\gamma\, hm\, B_0/2\pi. \tag{1.2}$$

The selection rule for the NMR transitions is that m can only change by one unit, $\Delta m = \pm 1$. Thus the transition energy is given as Eq. (1.3):

$$\Delta E = \gamma h B_0 / 2\pi \qquad (1.3)$$

Transitions between these two states can be induced by the electromagnetic radiation with frequency v (per second), which will effect such transitions as $\Delta E = hv$. Thus,

$$v = \gamma B_0 / 2\pi . \qquad (1.4)$$

Another equivalent expression is given by introducing the angular frequency ω_0 ($= 2\pi v$) (radian per second):

$$\omega_0 = \gamma B_0 . \qquad (1.5)$$

The spinning magnetic moment in the magnetic field undergoes precessional motion as a spinning top in the earth's gravitational field. The precessional frequency, called the Larmor frequency, is directly proportional to B_0 and μ, and is exactly equal to the frequency of electromagnetic radiation that induces a transition from one spin state to another (the resonance frequency). The resonance frequency depends both on B_0 and μ of the nucleus of interest. The natural abundance, γ, the relative receptivity, and the resonance frequencies of nuclei commonly employed in modern NMR spectroscopy are given in Table 1.1 [3, 4].

The relative populations of the lower-energy state, α, and the upper state, β, are given by Boltzmann's expression, expressed as the ratio of the numbers of α spin, N_α, and of β spin, N_β:

$$N_\alpha / N_\beta = \exp\left(2\mu B_0 / kT\right) = \exp\left(hv/kT\right) = 1 + hv/kT \text{ (as } hv \ll kT), \qquad (1.6)$$

where k is Boltzmann's constant and T is the absolute temperature. Even at a high field strength, $B_0 = 11.7$ T ($v = 500$ MHz for ^{1}H), obtained with a superconducting magnet, the excess population ($N_\alpha - N_\beta$) at 273 K is around 9 in 10^5 spins. This small difference in spin population is the reason for the low sensitivity of NMR compared with infrared or ultraviolet spectroscopy. However, the absorption coefficient is a constant for any nucleus, and thus the NMR signal intensity is directly proportional to the number of nuclei giving the signal, as far as the equilibrium population of spin states is maintained. This is an important characteristics of NMR spectroscopy. NMR spectra of polymer solutions are often measured at relatively high temperatures to achieve better spectral resolution. As the temperature of measurement is increased, however, N_α / N_β decreases, so the sensitivity lowers. Thus it should be noted that there is an optimum temperature for the measurement.

Table 1.1. Nuclear spinproperties for some common nuclei [3, 4]

Nucleus	I	$N/\%$[a]	$\gamma/10^7$ rad/T/s[b]	D[c]	NMR frequency at 2.3487 T (MHz)
^{1}H	1/2	99.99	26.75	6×10^3	100.0
^{2}H	1	0.015	4.11	8×10^{-3}	15.351
^{7}Li	3/2	92.58	10.40	2×10^3	38.862
^{10}B	3	19.58	2.87	2×10^1	10.747
^{11}B	3/2	80.42	8.58	8×10^2	32.084
^{13}C	1/2	1.11	6.73	1.0	25.144
^{14}N	1	99.63	1.93	6	7.224
^{15}N	1/2	0.37	−2.71	2×10^{-2}	10.133
^{17}O	5/2	0.037	−3.63	6×10^{-2}	13.557
^{19}F	1/2	100	25.18	5×10^3	94.077
^{23}Na	3/2	100	7.08	5×10^2	26.451
^{29}Si	1/2	4.70	−5.32	2	19.865
^{31}P	1/2	100	10.84	4×10^2	40.48
^{35}Cl	3/2	75.53	2.62	2×10^1	9.798
^{37}Cl	3/2	24.47	2.18	4	8.156
^{59}Co	7/2	100	6.32	2×10^3	23.63
^{117}Sn	1/2	7.61	−9.58	2×10^1	35.625
^{119}Sn	1/2	8.58	−10.02	3×10^1	37.27
^{207}Pb	1/2	22.6	5.54	1×10^1	20.92

[a] = Natural abundance; [b] = Gyromagnetic ratio; [c] = Receptivity relative to ^{13}C.

1.1.2 Relaxation

The various types of nonradiative transitions by which a nucleus in an upper spin state returns to a lower state are called relaxation processes. The processes may be divided into two categories, namely, spin–lattice relaxation and spin–spin relaxation. In spin–lattice relaxation, also called longitudinal relaxation, the energy of the spin system is converted into thermal energy of the molecular system containing the magnetic nuclei. The process is responsible for maintaining the unequal population of spin states. In spin–spin relaxation, also called transverse relaxation, a nucleus in its upper state transfers its energy to a neighboring nuclear spin. This relaxation process therefore does nothing about spin-state populations.

The time constants characterizing these relaxation processes are called the spin–lattice relaxation time, T_1[1], and the spin–spin relaxation time, T_2. Since the

---

[1] When a sample is put in the magnetic field, the induced magnetization does not come instantaneously into existence but builds up exponentially with time constant T_1. In flow NMR (Chap. 8) the samples flow into the magnetic field from outside zero field and the premagnetization process is required before acquiring NMR signals.

molecular motions of the neighboring nuclei in the same molecule are responsible for the relaxation processes, T_1 and T_2 measurements are used to study molecular motions of polymer chains (Chap. 7).

^{13}C NMR spectra are usually measured under the proton decoupling condition. One of the consequences of the decoupling is a change in the intensity of some peaks during decoupling, known as nuclear Overhauser enhancement (NOE) [9]. A quantitative derivation of the NOE for a two-spin system shows that when the relaxation is totally a dipole–dipole mechanism, the signal enhancement factor, η, depends on the gyromagnetic ratios of the nuclei A and X (γ_A and γ_X), as $\eta=0.5\gamma_A/\gamma_X$. The nondecoupled system has, by definition, an enhancement factor of 1 and therefore the decoupled system with the full NOE operating will have an intensity of $1+\eta$. The maximum enhancement due to NOE for ^{13}C with proton decoupling is 2.98.

For quantitative measurement, the NOE should be removed since the enhancement factor for each carbon signal might be different. For this purpose, the gated decoupling technique is employed, where the decoupler is turned on for a short time while the sample is pulsed and then turned off as the free induction decay (FID) (Sect. 1.2, Fig. 1.5) is acquired. (It is sometimes called inverse gated decoupling. The decoupling mode without NOE is termed NNE in Chap. 2.)

1.1.3 Chemical Shift

When a molecule containing the nuclei under observation is placed in a magnetic field, the electrons within the molecule shield the nuclei from the externally applied field. To gain an initial insight into the mechanism of the shielding, let us consider a hydrogen atom (Fig. 1.1).

In the presence of the magnetic field, the electron circulates in the direction shown (anticlockwise in the direction of B_0). The motion of the election is equivalent to an electric current flowing in a closed loop and as such it is associated with a secondary magnetic field which is also depicted in Fig. 1.1. The secondary

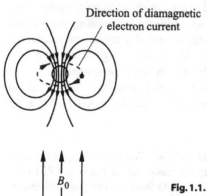

Fig. 1.1. Diamagnetic shielding of the nucleus of an isolated hydrogen atom in the presence of a magnetic field, B_0

magnetic field opposes the applied field, B_0, at the nucleus, which consequently experiences a total field which is slightly less than B_0. Thus the observed resonance frequency of a proton appears to be slightly less than that predicted from the value of B_0 and the gyromagnetic ratio of a proton. Equation (1.4) can be modified to Eq. (1.7), where σ is the shielding constant for the hydrogen atom:

$$\upsilon = \gamma B_0 (1-\sigma)/2\pi. \tag{1.7}$$

The equation tells us the resonance frequency υ is proportional to B_0 and thus the strength of the secondary magnetic field and consequently the chemical shift are likewise proportional to B_0. Therefore, the chemical shift, δ, is defined as the nuclear shielding divided by the applied field, and thus is only a function of the nucleus and its environment. It is always measured from a suitable reference compound (Sect. 1.4.2):

$$\delta = \frac{B_{ref} - B_{sample}}{B_{ref}} \times 10^6 \quad \text{(ppm)}, \tag{1.8}$$

where B_{ref} is the magnetic field at the reference nuclei and B_{sample} the field at the sample nuclei. Again it should be kept in mind that the chemical shift δ in parts per million (ppm) is a molecular parameter depending only on the measurement conditions (solvent, concentration, temperature) and not on the magnetic field or frequency applied for the spectrometer.

In ^1H and ^{13}C NMR spectroscopy, the recommended reference compound is tetramethylsilane (TMS). The recommended nomenclature is the δ scale, increasing in a downfield direction with the TMS peak being zero. Experimental determination of the chemical shift is described in Sect. 1.4.2.

1.1.4 Spin–Spin Coupling

Most ^1H NMR spectra consist not only of individual lines but also of groups of lines termed multiplets. The peak splitting into multiplet arises from nuclear interactions which cause splittings of energy levels and, hence, several transitions in place of the single transition expected otherwise. This type of interaction is called spin–spin coupling or scalar coupling.

Another mode of spin–spin coupling is direct dipolar coupling (through-space interaction of magnetic dipoles), which is proportional to $1/r^3$ (r is the distance between nuclei) and the gyromagnetic ratio of the nucleus, and also depends on the angle, θ, between r and B_0:

$$B_1 = \frac{\gamma h}{4\pi r^3} (3\cos^2 \theta - 1). \tag{1.9}$$

When the nuclei are in molecules that undergo rapid, random motion, as are molecules in solution, the dipolar interaction averages almost completely to zero.

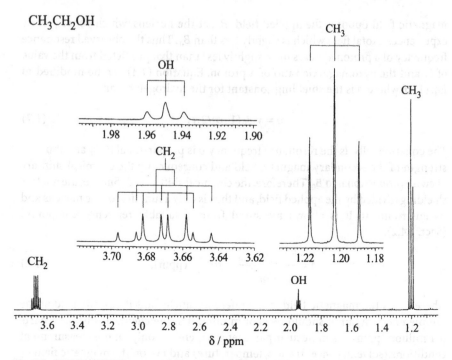

Fig. 1.2. 500 MHz ^{1}H NMR spectra of ethanol (5 wt/vol%) in CDCl$_3$ at 45 °C. 45° pulse, pulse repetition time 10 s, 16 scans

In the solid state the dipolar–dipolar interaction is not averaged out and the NMR signals are very broad. Nevertheless, high-resolution NMR spectra of solid samples are now obtainable by the combined use of high-power dipolar decoupling, magic-angle sample spinning, and cross-polarization methods. The advanced solid-state NMR spectroscopy of polymers has been covered by a recent publication [10].

The mechanism of scalar coupling has been suggested to involve the bonding electrons; information about the spin state of one nucleus A is transmitted to the other coupled one B via spin states of the bonding electrons. The energy level of spin A depends on the orientation of spin B and two spectral lines result, the difference in frequency being proportional to the energy of interaction or spin coupling between A and B. The magnitude of the interaction is given by the spin-spin coupling constant J_{AB} (expressed in hertz), and is independent of the applied field.

Spin–spin couplings observed in a ^{1}H NMR spectrum of highly purified ethanol in chloroform-d (CDCl$_3$) is shown in Fig. 1.2. Let us consider the methyl protons in relation to the possible spin arrangements of the two methylene protons. There are four possible arrangements for the methylene group. If we label the two protons A and B then we have (1) A and B both in parallel spin states, (2) A parallel and B antiparallel, (3) A antiparallel and B parallel, and (4) A and B both in antiparallel spin states (Fig. 1.3). Arrangements 2 and 3 are energetically equivalent. The magnetic

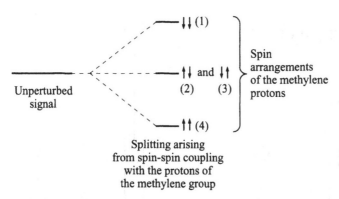

Fig. 1.3. The splitting of the signal from the methyl protons in ethanol by spin–spin interaction with the protons of the methylene group

effect of these arrangements is in some way transmitted to the methyl protons and so these protons will experience one of three effective fields according to the instantaneous spin arrangement of the methylene protons. Thus, for ethanol molecules in the whole sample there will be three equally spaced transition energies (frequencies) for the methyl protons. Since the probabilities of existence of each of the four spin arrangements are equal and arrangements 2 and 3 are equivalent, it follows that the intensities of the three transitions will be 1:2:1. The spacing of adjacent lines in the multiplets is a direct measure of the spin–spin coupling of the protons of the methylene group with those of the methyl group, and is known as the spin–spin coupling constant, J. The hydroxyl proton signal also shows three lines with different spacing from that observed for the methyl proton signal, indicating that the J values are different. The methylene proton signals are more complicated owing to the spin–spin couplings from the methyl and hydroxy protons (Fig. 1.2).

When a trace amount of acid is added to the ethanol, the spectrum changes as shown in Fig. 1.4. The lack of multiplicity of the hydroxyl proton signal is due to rapid exchange of hydroxyl protons, which is catalyzed by acid. As a result of this exchange any one hydroxyl proton, during a certain interval of time, is attached to a number of different ethanol molecules and thus experiences all possible spin arrangements of the methylene group. If the chemical exchange occurs with a frequency which is substantially greater than the frequency separation of the components of the multiplet from the hydroxyl proton signal, the magnetic effects corresponding to the three possible spin arrangements of the methylene protons are averaged out, and a single sharp absorption line is observed for the hydroxyl proton. In other words, rapid exchange causes spin decoupling between the hydroxyl and methylene protons. The quartet splitting of the methylene proton signal remains after the rapid exchange due to the spin coupling with methyl protons.

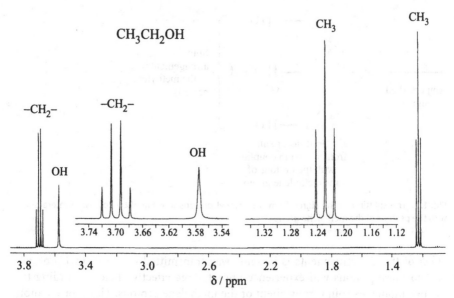

Fig. 1.4. 500 MHz ^{1}H NMR spectra of ethanol (5 wt/vol%) in CDCl$_3$ at 45 °C in the presence of a small amount of HCl. 45° pulse, pulse repetition time 10 s, 16 scans

1.2 Spectrometers

It is not our intention to discuss the details of NMR spectrometers but to give the general outline of them. High-resolution NMR spectrometers are categorized into continuous wave (CW) and pulsed Fourier transform (FT) spectrometers, both of which require a radio-frequency (RF) source and a magnetic field. The sample solution in a sample tube is placed in a probe which is set in the magnetic field. The sample tube is invariably rotated about its axis by using an air-flow turbine (usually 15–20 Hz) to average the magnetic field observed at each part of the sample about the spinning axis, producing increased resolution of the spectrum. The RF radiation is transmitted by a coil on the probe and detected either by the same coil (a single-coil spectrometer) or by a separate one (a cross-coil spectrometer). In the CW spectrometer, either the magnetic field or the RF is slowly varied to cover a certain range of chemical shift. When the resonance condition is satisfied, the sample absorbs energy from the RF radiation and the resulting signal is detected on the receiver coil, amplified, and recorded. In this method the frequency-domain spectrum is obtained directly.

In the pulse FT spectrometer, RF irradiation is made by a short high-energy RF pulse with a discrete frequency, f, known as the carrier frequency. When a short pulse with t seconds duration (pulse width) is turned on and off, we obtain not a single frequency but a range of frequencies centered at f with a bandwidth of roughly $1/t$; the frequency range covered is around $f \pm (1/t)$, (e.g., $t=10^{-5}$ s, range 10^5 Hz). Thus, by using a very short pulse with sufficient power, it is possible to

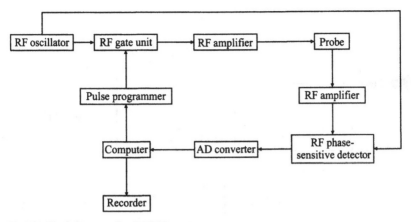

Fig. 1.5. Block diagram of an FT NMR spectrometer

excite or irradiate all the nuclei of a given species simultaneously. After the pulse is turned off, the spin system emits the energy, returning to thermal equilibrium of the spin states. The signal observed in this process is called the FID signal or simply FID, which is a time-domain spectrum. Fourier transformation of the FID gives the normal frequency domain NMR spectrum. A block diagram of an FT NMR spectrometer is shown in Fig. 1.5.

Two FIDs and their Fourier transformed signals from a single type of ^{1}H nucleus observed with a 270 MHz spectrometer are illustrated in Fig. 1.6. In Fig. 1.6a,

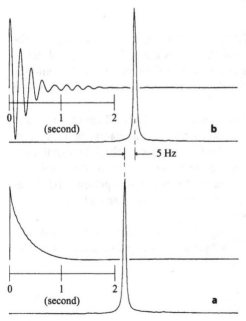

Fig. 1.6a,b. FIDs and their Fourier transformed spectra of H_2O (0.2 wt/vol%) in D_2O at 270 MHz and at 25 °C. 45° pulse, 1 scan

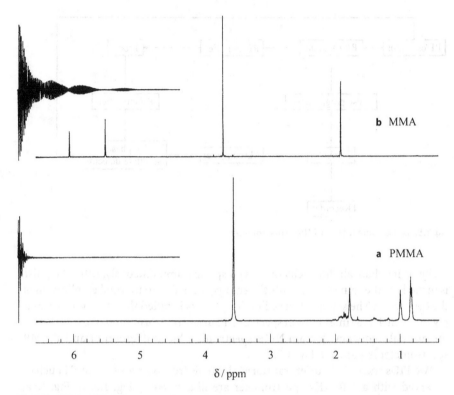

Fig. 1.7. 500 MHz ^{1}H NMR spectra of **a** PMMA (10 wt/vol%) in CDCl$_3$ at 55 °C and **b** MMA in CDCl$_3$ at 30 °C. 45° pulse, pulse repetition time 10 s, 16 scans

the observed frequency (270,167,557.3 Hz) is set to be exactly the same as the resonance frequency of the nucleus, and the FID is a simple-exponential decay. In Fig. 1.6b, the observed frequency is set 5 Hz larger than that in Fig. 1.6a, and the FID shows five waves per second and the Fourier transformed signal deviates 5 Hz from that in Fig. 1.6a.

Examples of FIDs and their Fourier transformed ^{1}H NMR spectra of poly-(methyl methacrylate) (PMMA) and its monomer, methyl methacrylate (MMA), are shown in Fig. 1.7a and b, respectively. The faster decay of the FID signal from PMMA than that from the monomer is due to faster relaxation of the spins or shorter relaxation times. The difference in the time-domain spectra (FID) is reflected in the frequency-domain spectra as a difference in peak widths; the faster the relaxation, the broader the peak width.

A pulse sequence representation of a simple FT NMR measurement is illustrated in Fig. 1.8. The effective strength of the RF pulse is usually denoted by the pulse angle or flip angle, α, which is defined as

$$\alpha = \frac{\omega_1}{2\pi}\tau = \frac{\gamma B_1}{2\pi}\tau \quad \text{(degree)}, \tag{1.10}$$

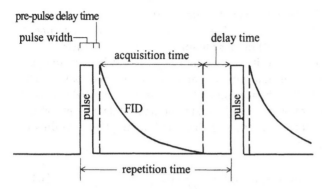

Fig. 1.8. Pulse sequence for FT NMR operation

where τ is the duration of the RF pulse (in seconds), ω_1 is the angular frequency (radians per second), γ is the gyromagnetic ratio, and B_1 is the amplitude of the RF field. To explain the pulse angle, let us consider a large number of spins, all possessing the same Larmor frequency (Sect. 1.1.1) with random distribution in their phase of rotation (Fig. 1.9a). The total magnetization of the sample is stationary and aligned along the B_0 direction (z-axis). The RF pulse is an oscillating magnetic field in the x–y plane, which can be regarded to be equivalent to two counterrotating magnetization vectors with Larmor frequency (Fig. 1.9b). If the rotating set of coordinates, called a "rotating frame", is chosen to rotate at the same speed and in the same direction as the rotating RF field, the RF field appears to be fixed to one axis (x'-axis in Fig. 1.9c). As electromagnetic theory tells us, the interaction between the total magnetic moment, M_0, along the z-axis and the magnetic field B_1 along the x-axis causes the M_0 to precess about B_1 in the rotating frame. When the RF pulse B_1 is applied for τ seconds, the total magnetization vector M_0

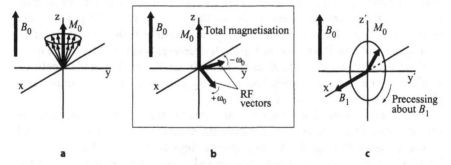

Fig. 1.9a–c. Behavior of magnetic moments and total magnetization. **a** Total magnetizations, M_0, aligned along the z-axis (B_0). **b** Applied RF electromagnetic field perpendicular to B_0 may be regarded as two oppositely rotating RF vectors with angular frequency of $\pm\, \omega_0$. **c** "Rotating frame of reference", which rotates with an angular frequency of $\pm\, \omega_0$, gives an apparently static magnetic field B_1 perpendicular to B_0. M_0 precesses about B_1

in the rotating frame is tilted from its equilibrium position along B_1 by an angle α (degree), which is called the "pulse angle" (Fig. 1.9c). For a single-pulse experiment, the maximum signal-to-noise (S/N) ratio is obtained by using a 90° pulse.

The acquisition of the FID and its storage in a digital form is normally performed automatically by means of a computer attached to the spectrometer. The time required to acquire the FID, the acquisition time, is determined by the spectral width, Δ (frequency range of observation), and the number of data points, N, the number of computer memory locations available for data sampling according to the Nyquist theorem (Eq. 1.11):

$$\text{Acquisition time} = \frac{N}{2\Delta} \text{ (s).} \qquad (1.11)$$

With N data points, the transformed spectrum contains only $0.5N$ real data points. Thus, the highest precision with which a single value of the chemical shift measurement can be determined is called digital resolution and is given as

$$\text{Digital resolution} = 2\Delta/N \text{ (Hz).} \qquad (1.12)$$

^{1}H NMR spectra are usually measured with a spectral width of 10 ppm, which, for example, corresponds to 5,000 Hz when the spectra are measured at 500 MHz. When the number of data points is 64k ($2^{16} = 65,536$), the acquisition time required is 6.55 s and the digital resolution is 0.1526 Hz (Sect. 2.1).

When a signal exists outside the frequency range, a folded signal will appear in the transformed spectrum, owing to the limited number of data points. The foldback signal can be distinguished from real signals by changing either the frequency range or the carrier frequency. The ^{13}C NMR spectra of MMA measured over two different frequency ranges are shown in Fig. 1.10; one covers whole signals and the other, with a narrower range, exhibits folded signals.

The time interval between two successive RF pulses, roughly equal to the sum of the acquisition time and the delay time, is called the pulse repetition time or pulse interval (Fig. 1.8). To obtain an accurate signal intensity, it must be sufficiently long for the magnetization to relax to the equilibrium state. With a 90° pulse, a pulse repetition time of $5T_1$ allows the magnetization to recover to 99.3% of the equilibrium value along the B_0 axis. As the flip angle decreases, less time is required to reestablish equilibrium along the B_0 axis. The optimum flip angle and the repetition time for a better S/N ratio depend on T_1 and T_2, and are usually determined empirically as described later (Sect. 2.1).

The dynamic range of an FT spectrometer, that is, the ratio of the largest to the smallest peaks in the spectrum, depends on the number of bits (binary digits) used for digitizing the FID signal in the analogue-to-digital converter (ADC). With a 12-bit ADC the dynamic range must not exceed 4,096:1, and with a 16-bit ADC 65,536:1. Experimental elucidation of the dynamic range is given in Sect. 2.4.

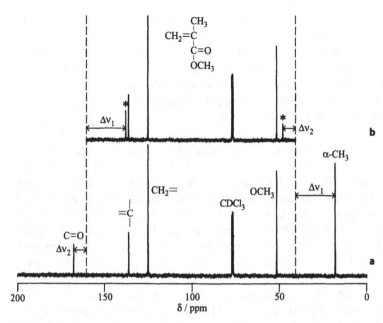

Fig. 1.10. Folding of peaks from outside the sampling frequency range in the 100 MHz ^{13}C NMR spectrum of MMA. 10 wt/vol% CDCl$_3$ solution, 45° pulse, pulse repetition time 3 s, 64 scans, spectral width **a** 20,000 Hz without folding peaks, **b** 12,000 Hz with folding peaks

1.3 Observable Nuclei

The most frequently and widely utilized NMR nucleus is ^{1}H owing to its high natural abundance and high sensitivity. In addition to the ^{1}H nucleus, the ^{13}C nucleus has been routinely utilized in NMR analysis of polymers [11], in particular after FT NMR was introduced. The chemical shift for usual polymers ranges around 200 ppm, much larger than for ^{1}H NMR, and provides us with a greater possibility of observing more detailed structural differences. ^{13}C NMR spectra are usually obtained under ^{1}H decoupling conditions (Sect. 1.1.2). Freed from large C–H spin coupling (around 100–150 Hz), the spectra become much simpler and, at the same time, apparent S/N ratios of individual signals become larger owing to the lack of splitting and to the enhancement through NOE. The low natural abundance of ^{13}C nuclei results in a low possibility of observing ^{13}C–^{13}C spin–spin coupling. This makes the spectra simpler but information on C–C connectivity in the chains is scarcely obtainable.

Besides ^{1}H and ^{13}C nuclei, several other NMR observable nuclei have been utilized in polymer analysis. Because ^{15}N, ^{19}F, ^{29}Si, and ^{31}P nuclei are found in several important classes of polymers, their NMR characteristics are outlined in the following.

Fluorine-19 has several attributes which make it suitable as an NMR label for studies of biological systems [12, 13], ^{19}F has I=1/2 and 100% natural abundance, and its NMR sensitivity is only slightly smaller than that for ^{1}H. ^{19}F chemical shifts span a much larger range than those of ^{1}H; approximately 300 ppm, and hence a slight structural difference may cause a readily discernible chemical shift difference. Large coupling constants for ^{19}F–^{19}F and ^{19}F–^{1}H also provide important structural information. In addition to these favorable NMR properties, the fluorine atom is also a good label from the viewpoint of its chemical properties, since its size is rather small and carbon–fluorine bonds are very inert. Thus, the labeling by fluorine atoms might cause only minor perturbations, which is particularly important in the study of biologically active molecules. In the field of synthetic polymers, fluorinated initiators have been used to study the end-group structures of the polymer obtained therewith. For example, p-fluorobenzoyl peroxide was used for the radical polymerization of styrene and ^{19}F NMR signals of the initiator fragments were assigned to the benzoyl and phenyl end groups [14]. Of course, ^{19}F NMR is also inevitable for the structural analysis of fluoropolymers. ^{19}F chemical shifts of fluoropolymers are much more sensitive to their microstructures than their ^{13}C chemical shifts are. ^{19}F NMR is therefore the method of choice for studying fluoropolymers [15, 16].

^{31}P also has 100% natural abundance and a large gyromagnetic ratio, and its NMR is easily observed as ^{1}H nuclei. The ^{31}P chemical shift ranges over 600 ppm and is sensitive to a variety of structural factors, such as the oxidation state of phosphorus, substituents, and steric structure [17]. ^{31}P NMR has been utilized for studies of reactive intermediates which undergo changes between at least two oxidation states (Sect. 5.1). 2-Phenylphospholane undergoes ring-opening polymerization by cationic initiators. The reactive intermediate of the polymerization may be either a phosphonium ion or covalent species, depending on the type of initiator. The polymerization with trifluoromethane sulfonate initiator gives the phosphonium-type species which shows a ^{31}P NMR signal at 101 ppm, while that with iodomethane initiator does not show this type of signal but a signal at 37 ppm attributable to a phosphine oxide type unit with a 3-iodopropyl end group [18]. Thus, ^{31}P NMR is a quite sensitive tool to investigate polymerization mechanisms involving phosphorus-containing species.

The ^{29}Si nucleus is a valuable probe for the analysis of silicon-containing organic and inorganic polymers. Though more abundant than ^{13}C (natural abundance 4.7 for ^{29}Si versus 1.1% for ^{13}C), the ^{29}Si nucleus has a smaller nuclear dipole moment and gyromagnetic ratio than ^{13}C, and thus ^{29}Si NMR is only about twice as sensitive as ^{13}C NMR. When broadband proton decoupling is used to remove ^{29}Si–^{1}H scalar coupling, the ^{29}Si signal is often reduced in intensity differently from ^{13}C NMR, because its γ value is negative. Moreover, the spin–lattice relaxation times for ^{29}Si nuclei in solution are long, much like those of ^{13}C nuclei, and a rather long pulse-repetition time is required.

The inherent sensitivity of the ^{15}N nucleus and its natural abundance are substantially less than those of ^{13}C. However, pulsed FT NMR techniques have made the ^{15}N nucleus observable at natural abundance. ^{15}N NMR spectroscopy is most often performed on enriched samples, particularly biologically important polymers, such as polypeptides and nucleic acids [19, 20].

^{14}N, the most abundant nucleus of nitrogen, is also an NMR observable nucleus, but because of its spin (I=1), quadrupolar relaxation makes the NMR signal broad and thus ^{14}N NMR has not been well appreciated in polymer analysis.

The deuterium-labeling method in ^{1}H NMR serves as a means of removing spin-coupling of a specific part of a molecule. Examples in polymer analysis are stereochemical sequence analysis of vinyl polymers of α- or β,β-deuterated vinyl monomers. Reasonable availability of deuterated compounds from commercial sources helps the use of deuterated compounds in polymer science, such as deuterated initiators and monomers. The natural abundance of ^{2}H (or D) is quite low (0.015%), so the NMR signals for ^{2}H-enriched groups in a polymer are hardly affected by the background signals owing to the unenriched part of the polymer molecules. A problem of ^{2}H NMR is that the signals cannot be used for NMR locking during the measurement, since ^{2}H signals usually from deuterated solvents are used to stabilize the fluctuation of the magnetic field strength (NMR locking). Thus accumulation of ^{2}H NMR should be made without NMR locking so that accumulation for a long time is not appropriate. Nevertheless, the short spin–lattice relaxation time of the ^{2}H nucleus allows a short duration between the observed pulses and spectra with an acceptable S/N ratio can be obtained in a limited time. The accuracy of the intensity measurement in ^{2}H NMR for end-group analysis of polymers has been well established by using a PMMA sample having a $CH_3(CD_3)_2C-$ group by comparing the results with those of ^{1}H NMR analysis for the CH_3 signal of the t-butyl end group [21].

1.4 Sample Preparation

Sample preparation in high-resolution NMR experiments for polymers involves dissolving a polymer with high molecular weight in an appropriate solvent and introducing the viscous polymer solution in an NMR sample tube. The important points are the selection of an appropriate solvent and the handling of a viscous solution of polymer with the removal of suspended dust in the solution. In this section the techniques and problems relating to the sample preparation are described. A few minutes spent thinking of the problems of sample preparation can save hours of wasted time afterwards.

1.4.1 Solvent for NMR Measurement

In ^{1}H NMR and ^{13}C NMR experiments the solvent signals will obscure certain regions of the spectrum and totally or partially deuterated solvents are usually used to avoid this complication. The intensities of the ^{13}C NMR signals of deuterated solvents are greatly reduced owing to the loss of NOE from protons in addition to the splitting into several lines by deuterium coupling. The boiling and melting points and the ^{1}H, ^{2}H, and ^{13}C chemical shifts of deuterated solvents commonly used in NMR experiments are shown in Table 1.2. The boiling and melting points

Table 1.2. Boiling Point (b.p.) and melting point (m.p.) and ^{1}H, ^{2}H, and ^{13}C chemical shifts (ppm) of commonly used deuterated NMR solvents (ppm) at 27 °C. Observed frequency: ^{1}H 500 MHz, ^{2}H 41.3 MHz, ^{13}C 125 MHz. Chemical shift standard: ^{1}H TMS,1 wt/vol%; ^{2}H TMS, 3 wt/vol% (signal of ^{2}H in natural abundance); ^{13}C TMS, 1 wt/vol%. When TMS-d_{12} is used as a chemical shift standard, the ^{2}H chemical shift values are slightly different, for example, the shift values for chloroform-d is 7.279 ppm from TMS-d_{12} and 7.229 ppm from TMS-d_1

Solvent		^{1}H	^{2}H	^{13}C	b.p. (°C)	m.p. (°C)
Acetic acid	Methyl	2.035	2.007	20.03	118	16.7
	Carboxyl	11.567	11.459	178.40		
Acetone	Methyl	2.051	2.031	29.83	56.5	−94
	Carbonyl	–	–	206.70		
Acetonitrile	Methyl	1.938	1.913	1.35	81.6	−45
	Nitrile	–	–	118.34		
Carbon tetrachloride		–	–	95.96	76.7	−23
Chloroform		7.258	7.229	77.01	61.6	−63.5
Cyclohexane		1.382	1.360	26.37	80.7	6.5
Deuterium oxide[a]		4.757	4.753	–	101.4	3.8
Dichloromethane		5.317	5.284	53.83	39.8	−95
N,N-Dimethyl-	Methyl	2.743	2.729	30.11	153	−61
formamide	Methyl	2.914	2.893	35.21		
	Aldehyde	8.022	8.021	162.70		
DMSO		2.500	2.493[b]	39.46	189	18.5
Dioxane		3.529	3.532	66.49	101.1	11.8
Ethanol	Methyl	1.114	1.104	17.23	78.5	−130.0
	Methylene	3.560	3.541	56.85		
Methanol		3.306	3.294	49.05	64.7	−97.8
Nitromethane		4.337	4.312	62.85	101.2	−29
Tetrahydrofuran	Methylene	1.723	1.707	25.32	66	−108.5
	Methylene	3.576	3.567	67.40		
Trifluoroethanol	Trifluoro-methyl	–	–	126.19	77.8	−43.5
	Methylene	3.898	3.869	61.48		
	Hydroxy	5.136	5.080	–		
Benzene		7.156	7.185	128.04	80.1	5.5
o-Dichlorobenzene	C1, 2	–	–	132.43	180.5	−17.0
	C3, 6	7.193	7.229	130.04		
	C4, 5	6.935	6.973	127.18		
Nitrobenzene	C1	–	–	148.56	210	6
	C2, 6	7.511	7.523	123.46		
	C3, 5	8.122	8.139	129.48		
	C4	7.678	7.688	134.83		

Table 1.2 (continued)

Solvent		¹H	²H	¹³C	b.p. (°C)	m.p. (°C)
Pyridine	C2, 6	8.728	8.727	149.92	115	−42
	C3, 5	7.207	7.223	123.52		
	C4	7.575	7.591	135.52		
Toluene	Methyl	2.087	2.081	20.43	110.6	−95
	C1	–	–	137.51		
	C2, 6	7.094	7.117	128.87		
	C3, 5	6.975	7.002	127.98		
	C4	7.013	7.035	125.15		

ª Chemical shift standard TSP-d_4.
ᵇ The chemical shift refers to acetone-d_6 (2.051 ppm).

are for the corresponding nondeuterated solvents, but they may be little different from those of the deuterated solvents. Figures 1.11 and 1.12 show 500 MHz ¹H NMR and 125 MHz ¹³C NMR spectra of commonly used deuterated NMR solvents, respectively. In most of the spectra the expanded signals of the remaining protons or of the carbons bearing the remaining protons are also shown. These may be a good aid for the selection of the solvent.

As mentioned in the previous section, in most of the modern NMR instruments the deuterium signal from the solvent is used for an NMR lock system to avoid fluctuation of the magnetic field strength. In the case where a nondeuterated solvent is used, it should contain a small amount of deuterated solvent to provide a ²H NMR signal with enough intensity for the lock system. If the deuterated solvent is immiscible with the nondeuterated solvent, the former may be placed in the central capillary of a coaxial tubing (Fig. 1.18b).

Recent advances in NMR spectrometers provide us with spectra with high resolutions and high S/N ratios. This enables us to take the spectrum of the sample solution at very low concentrations. In the case of ¹H NMR measurement, the singlet signal due to acetone at a concentration of 1.1×10^{-5} g hydrogen/l can be easily detected with a 500 MHz NMR spectrometer [22]. One tenth of a microgram of acetone (hydrogen content of acetone 10.3 wt%) in 0.5 ml $CDCl_3$ was enough to be detected by 500 MHz ¹H NMR using a 5-mm NMR sample tube (sample volume 0.5 ml). Even in the case of 125 MHz ¹³C NMR measurement, 0.6 mg ethylbenzene in 0.6 ml $CDCl_3$ gave the ¹³C NMR signals shown in Fig. 1.13 when measured with 18,000 accumulation and 45° pulses for 15 h.

Measurements of spectra of the solution at a low concentration inevitably require solvents with high purities since the signals of the impurities obscure the signals of the sample of interest. This is a matter of concern particularly in the cases of polymers, which usually show much broader signals with low S/N ratios compared with those of low molecular weight compounds.

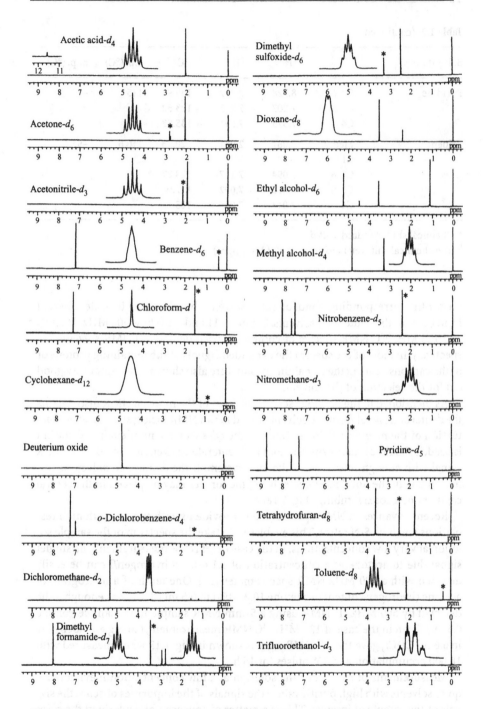

Fig. 1.11. 500 MHz ^{1}H NMR spectra of commonly used deuterated NMR solvents measured at room temperature. The peak marked with an *asterisk* is due to H_2O and/or HDO. 45° pulse, pulse repetition time 20 s

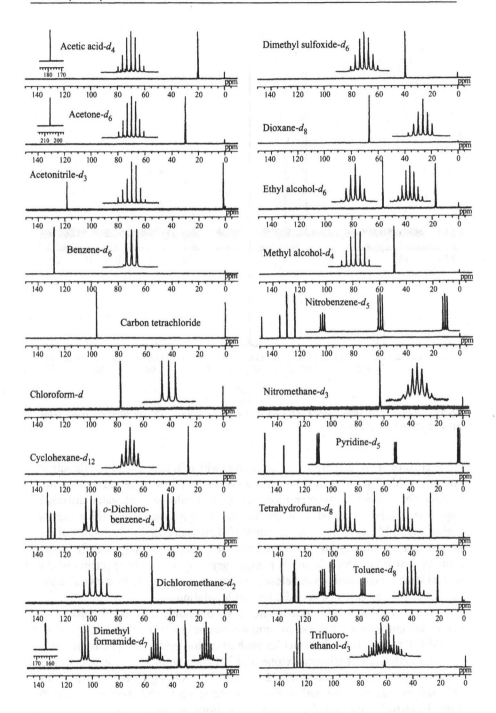

Fig. 1.12. 125 MHz ^{13}C NMR spectra of commonly used deuterated NMR solvents measured at room temperature. 45° pulse, pulse repetition time 10 s

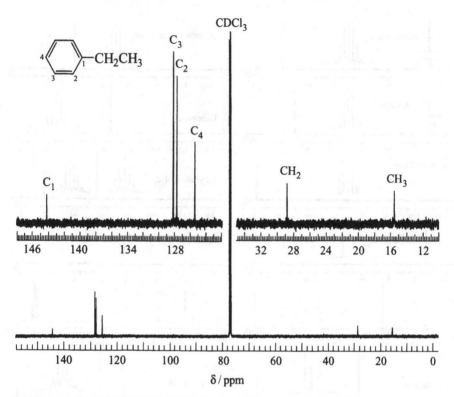

Fig. 1.13. 125 MHz ^{13}C NMR spectra of 0.1 wt/vol% ethylbenzene in CDCl$_3$ at 30 °C. 45° pulse, pulse repetition time 3 s, 18,000 scans

In ^{1}H NMR spectra, there are three sources which interfere with the signals of interest: residual protons in deuterated solvents, dissolved water, and other impurities. The amounts of the corresponding nondeuterated solvent, water, and other impurities can be measured by an absolute intensity measurement with a coaxial tubing method [23, 24] (Sect. 2.3). A survey [22] revealed that the amounts of the nondeuterated solvents are usually within the guaranteed range but the water contents depend on the solvents. It is necessary to measure the spectrum of the solvent itself for contamination before running the spectrum of your own sample, particularly in the case of a very dilute sample solution.

Deuterated chloroform is one of the most commonly used NMR solvents. A deuterated chloroform that was in common use in our laboratory was found by ^{1}H NMR to contain several impurities, such as trichloroethane, dichloromethane, dibromomethane, acetonitrile, acetone, and TMS in amounts of a few ppm or less besides chloroform and water (Fig. 2.11, Sect. 2.4). Among these impurities the haloalkanes may be contained originally in the solvent, while acetone and TMS might be introduced during laboratory use and/or storage.

Water is also introduced into the solvent during storage and its content increases gradually with time. For example, as described in Sect. 2.4, the water content in a

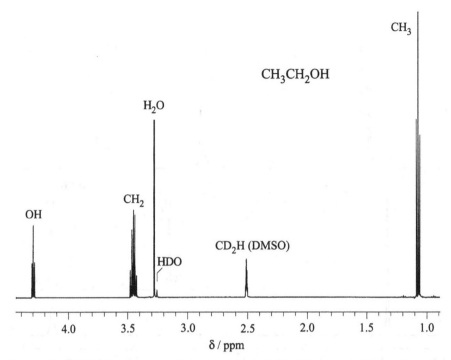

Fig. 1.14. 500 MHz ^{1}H NMR spectra of ethanol (0.35 wt/vol%) in DMSO-d_6 at 30 °C. 90° pulse, pulse repetition time 25 s, 16 scans

deuterated acetone solution placed in an NMR sample tube with a plastic cap increases at the rate of 100 ppm/day when the tube is kept in an ordinary chemistry laboratory. The rate of water uptake depends strongly on the storage conditions. When stored in a refrigerator, the rate of water uptake is greatly reduced (5 ppm/day). Sealing the tube with a shrinkable polyethylene tube (Fig. 1.22) effectively decreases the uptake rate to about 10 ppm/day in the laboratory (Sect. 2.4).

Some of the deuterated solvents contain D_2O, DHO, as well as H_2O; only the last two show NMR signals in the ^{1}H NMR spectrum (Fig. 2.12, Sect. 2.4). The water in the solvent will interfere with the observation of the signals of other exchangeable protons in the sample of interest. In the presence of D_2O, the signals of exchangeable protons of samples may disappear owing to the D–H exchange. In dimethyl sulfoxide (DMSO), the signals of water and of other exchangeable protons can be distinguished as shown in Fig. 1.14, owing to the strong solvation of water by DMSO.

The overlap of the water signal with the signals of the sample of interest can be avoided by changing the temperature of measurement; an increase in the temperature shifts the water signal upfield and a decrease shifts it downfield. Addition of a small amount of aqueous solution of deuterium chloride (DCl/D_2O) shifts the water signal remarkably downfield. So this is very effective for avoiding the interference by the water signal if DCl is not harmful to the sample of interest. Even in the

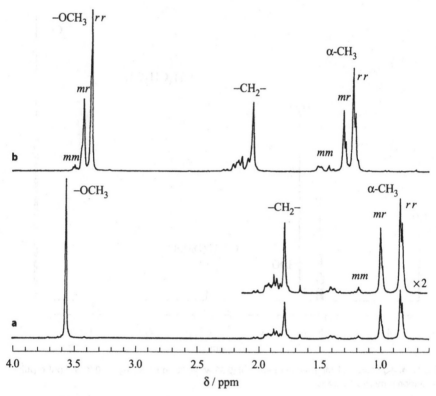

Fig. 1.15. 500 MHz ¹H NMR spectra of radically prepared PMMA (10 wt/vol%) measured in **a** chloroform-*d* at 55 °C and in **b** benzene-*d₆* at 80 °C. 45° pulse, pulse repetition time 10 s, 16 scans

case where the sample solution is immiscible with DCl/D₂O, the method is sufficiently effective.

Sample solutions of polymers sometimes contain small amounts of low molecular weight impurities, including unreacted monomer and the solvent used for polymerization or purification. The signals of these low molecular weight impurities are rather easily distinguished from the polymer signals by checking the spin–lattice relaxation times (T_1) for the signals; the signals of low molecular weight impurities show much larger values of T_1 than those of polymers. The T_1 values for the signals of end groups of polymers are larger than those for the signals of in-chain groups, but are smaller than those for low molecular weight impurities.

Samples obtained from high performance liquid chromatography (HPLC) or size exclusion chromatography (SEC) fractionation sometimes contain a relatively large amount of impurities due to the eluant and/or the extracts from the column packing materials. In such cases it is recommended to take the NMR spectrum of the evaporated residue of the eluant from the chromatograph and to use it to make the difference spectrum.

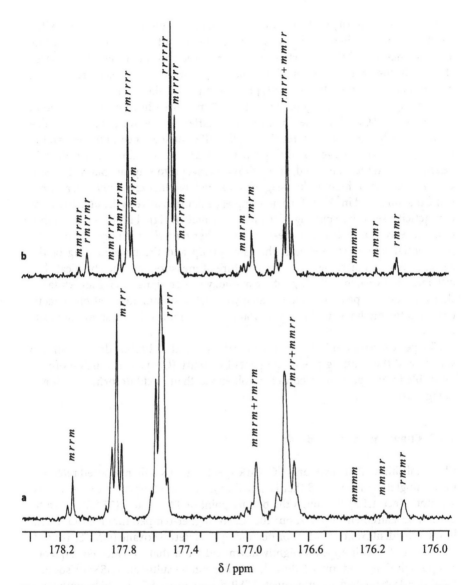

Fig. 1.16. 125 MHz carbonyl carbon NMR spectra of radically prepared PMMA (10 wt/vol%) measured in **a** chloroform-*d* at 55 °C and in **b** toluene-*d*$_8$ at 105 °C. 45° pulse, pulse repetition time 3 s, 18,000 scans

Beyond the impurity problems, the solvent used may influence the quality of the spectrum obtained in several ways, including the resolution, the peak separation, and the chemical shifts of certain peaks. Aromatic solvents usually cause large changes in the chemical shifts of the solute compared with the spectra taken in nonaromatic solvents. One of the typical examples is shown in Fig. 1.15 of the ^{1}H NMR spectra of radically prepared PMMA measured in chloroform-d and in benzene-d_6 at 500 MHz. The chemical shift difference between the peaks due to the α-methyl groups in stereochemically different sequences, mm, mr, and rr triads (Chap. 3), is larger in the spectrum measured in chloroform-d than in benzene-d_6. On the other hand, the methoxy methyl proton signal obtained in benzene-d_6 splits into three peaks assignable to the triad tacticity (mm, mr, and rr), while that obtained in chloroform is a singlet peak insensitive to the tacticity. Similar tacticity-sensitive splitting of the methoxy methyl signal is observed in other aromatic solvents, such as toluene-d_8 and nitrobenzene-d_5, and may be due to the preferential interaction between the ester group and the benzene ring of the solvents. With an increase in the temperature of measurement the chemical shift difference between the splittings of the methoxy proton signal decreases, while the sharpness of each peak is enhanced. So if you wish to obtain information on tacticity from the methoxy methyl signal, you have to compromise these two opposite factors.

The peak separation in the carbonyl carbon signal of PMMA depends on the solvent and the splitting due to the pentad tacticity (Chap. 3) is much more remarkable in the spectrum recorded in toluene-d_8 than in chloroform-d as shown in Fig. 1.16.

1.4.2 Chemical Shift Standards

Chemical shifts in ^{1}H NMR and ^{13}C NMR spectra are usually measured referring to the singlet signal of TMS that is added in a very small amount to the sample solution (Sect. 1.1.3). Although the boiling point of TMS is low (27 °C), it can be used in measurements up to about 100 °C. For higher-temperature measurement, hexamethyldisiloxane (HMDS) or octamethylcyclotetrasiloxane (OMTS) is used, the signal of which appears slightly downfield from that of TMS. For aqueous sample solutions, sodium 4,4-dimethyl-4-silapentane sulfonate (DSS) or sodium [2,2,3,3-d_4]3-trimethylsilylpropanoate (TSP-d_4) can be used. It should be noted that DSS shows weak signals due to the three methylene groups as well as a strong peak due to three methyl groups. When these ionic compounds are not appropriate for aqueous sample solutions as standards, the use of 1,4-dioxane is recommended. When the standard reagent is insoluble in the sample solution, a capillary containing the standard (external standard) is put in the sample solution.[2]

The precision coaxial tubing (Fig. 1.18b, Sect. 1.4.3) is useful for this purpose; the standard reagent is placed in the central capillary to avoid the occurrence of a

---

[2] Foodnote s. page 25.

Table 1.3. Chemical shift standards for NMR measurements

Compound	Chemical formula	b.p. (°C)	δ (ppm)[a]
TMS	$(CH_3)_4Si$	27	0 (0)
HMDS	$[(CH_3)_3Si]_2O$	100	0.065 (1.94)
OMTS	$[(CH_3)_2SiO]_4$	175	0.093 (0.75)
DSS	$(CH_3)_3Si(CH_2)_3SO_3Na$		0.017 (0.15)[b]
TSP-d_4	$(CH_3)_3Si(CD_2)_2COONa$		0 (0)[b]
1,4-Dioxane	$[O(CH_2)_2]_2$	101	3.764 (69.41)[b]

[a] The values in parentheses represent the ^{13}C NMR chemical shift.
[b] The values in aqueous solution.

strong spinning sideband. The chemical shift standards for the NMR of ^1H, ^{13}C, and several other nuclei are summarized in Table 1.3.

Sometimes the chemical shifts are measured referring to the residual proton signals of deuterated solvents and converted into a TMS scale using the literature values for the solvents as those in Table 1.2. In such a case an important point that should be noted is that the chemical shifts of the residual protons in some deuterated solvents depend on the kind of solutes and their concentrations. For example, the ^1H NMR signal of $CHCl_3$ in $CDCl_3$ shifts downfield when ester, ketone, ether, alcohol, or carboxylic acid is dissolved in it, and shifts upfield when aromatic or aliphatic hydrocarbon is dissolved. Haloalkanes as solutes in $CDCl_3$ do not affect the chemical shift of $CHCl_3$ as shown in Fig. 1.17. It is clear from the figure that the solute-dependent shift of $CHCl_3$ is within ±0.01 ppm when the concentration of solute is less than 0.1 mol/l. The $CHCl_3$ signal in $CDCl_3$ solution is often used as an internal standard for chemical shift. When higher accuracy of the chemical shift measurement (e.g., 0.001 ppm or less) is required, however, the solute-dependent shift of $CHCl_3$ should be considered; at least the sample concentrations are to be kept constant.

---

[2] If an interface exists between the sample and reference, there is a difference in the magnetic field between them given by

$$(B_0)_s - (B_0)_R = B_0 [4/3\pi (\chi_R - \chi_S)]$$

for the usual cylindrical sample tube under a vertical magnetic field as in a superconducting magnet (χ_S and χ_R are the volume magnetic susceptibilities of the sample and the reference, and are negative for ordinary diamagnetic materials). The true chemical shift is given by

$$\delta (true) = \delta (observed) - 4/3\pi \times 10^6 (\chi_R - \chi_S).$$

In contrast, if a reference compound exists in the sample solution, it is termed an "internal reference".

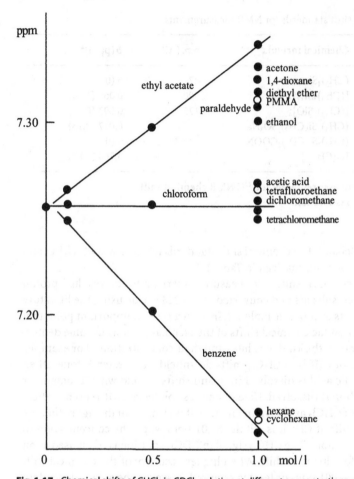

Fig. 1.17. Chemical shifts of CHCl$_3$ in CDCl$_3$ solution at different concentrations of various solutes [25]

A small amount of silicone grease happens to intermingle in the sample solution during preparation of the solution or the sample itself. The ^{1}H NMR and ^{13}C NMR signals of silicone grease appear very close to those of TMS and must be carefully distinguished from the TMS signal. In particular cases such as silyl esters, metal hydrides, and organometallic compounds, the signal from the standard TMS is sometimes obscured by the signals of the sample of interest. In such a case addition of TMS should be avoided and a solvent signal may be used as chemical shift reference.

1.4.3 NMR Sample Tube and Sample Manipulation

For ordinary NMR measurements, the sample tube of type A in Fig. 1.18 with an outer diameter of 5 mm is used. Tubes with low cylindrical symmetry often pro-

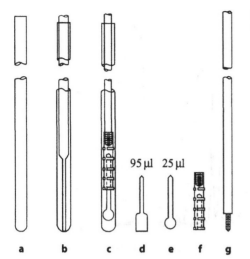

Fig. 1.18. Several kinds of NMR sample tubes [25]

95 μl 25 μl

a   b   c   d   e   f   g

vide spectra with strong spinning sidebands. Generally, the higher the magnetic field is, the more tightly specified tubes should be used.

A wide range of sample tubes with different specification are offered by manufacturers. The coaxial tubing B in Fig. 1.18 was originally designed for external referencing of the chemical shift; the chemical shift standard is placed in the central coaxial capillary and the sample solution is in the surrounding annulus (Sect. 1.4.2). It can also be used for absolute quantitative analysis including microanalysis and NMR determination of magnetic susceptibility of organic compounds (Sects. 2.3–2.5). When the amount of sample available is limited, use of a microcell is favored to minimize the volume of the NMR solvent so that signals due to the solvent itself and impurities in the solvent can be minimized. The coaxial tube can be used as a microcell for measuring the spectrum of small samples; the sample solution is placed in a central capillary (about 0.03 ml). The ^{1}H NMR spectra of 0.114 mg radically prepared PMMA measured in benzene-d_6 by using the coaxial tubing and the ordinary 5-mm tube are shown in Fig. 1.19 [25]. The signals due to the impurities in the solvent are much weaker in the spectrum measured in the capillary and so the quality of the spectrum is much improved.

Loading the sample solution in the central capillary and cleaning the inside of the capillary is easily done using a small syringe or a microsyringe with a long thin needle. One of the advantages in using the coaxial tube as a microcell is that high spectral resolution is more easily obtained compared with the microcells C and D in Fig. 1.18.

The microcell shown in Fig. 1.18c is the modified Frei–Niklaus microcell [26], in which a short-stemmed bulb is supported in a standard sample tube with a friction-grip Teflon chuck. The sample solution is loaded in the bulb with a microsyringe or a small hypodermic syringe. The space around the spherical cell is filled with nonprotic liquids. This minimizes the wobbling of the spherical cell. The

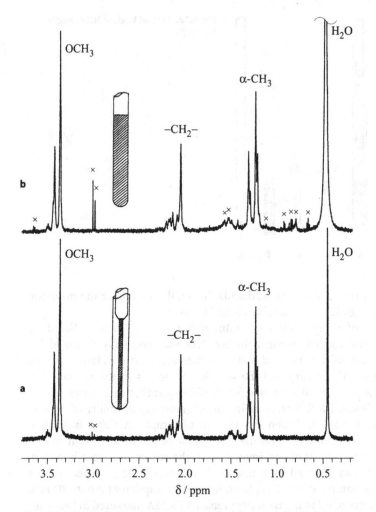

Fig. 1.19. 500 MHz ^{1}H NMR spectra of 0.114 mg radically prepared PMMA measured in benzene-d_6 by using **a** a coaxial tube and **b** an ordinary 5-mm NMR tube under the same conditions [25]. 45° pulse, pulse repetition time 3 s, 512 scans, 55 °C

sample cell and chuck can be easily positioned in the NMR tube with the aid of a removable rod for optimum coupling of the sample to the receiver coil. The chemical shift standard and the deuterated compound for NMR lock should be contained in the sample solution. The disadvantage of this microcell assembly is that a high resolution of the spectrum is rather difficult to obtain with superconducting magnet spectrometer. The problem of the obtainable resolution can be greatly improved by use of the oblong bulb shown in Fig. 1.18d.

It should be noted that the microcell equipped with a spherical cell (Fig. 1.18c) can be conveniently used for the determination of magnetic susceptibility with an iron magnet spectrometer [27] and that the microcell equipped with the oblong cell

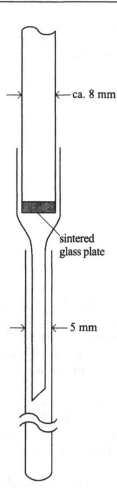

Fig. 1.20. Filtration assembly of a short glass tube with a sintered glass plate at one end [25]

←— ca. 8 mm

sintered glass plate

←— 5 mm

is useful for the determination of the spin–lattice relaxation time of polymers at high magnetic fields and high temperatures (Sect. 7.3).

It is necessary to keep the NMR sample cells clean and free from dust and scratches. One of the recommended procedures to clean the cells used for polymer solutions is to wash them with good solvents for the polymers first to evacuate most of the polymer samples. Then, the tube is rinsed with a nonsolvent for the polymer (e.g., methanol) to remove a very small amount of the polymer residue from the cell wall and finally with distilled water. Soaking in a proprietary degreasing solution followed by a water rinse is sometimes necessary after almost completely rinsing out the polymer. The cleaned tube is then dried in an oven while avoiding the opportunity for dust to enter. For complete drying, the tube is blown by filtered and dried nitrogen gas using a long thin needle of a syringe while it is heated by an appropriate means, such as a heat gun.

High-resolution spectra can only be obtained for solutions which are free from suspended dust. The best way to achieve this is to filter the sample solutions directly

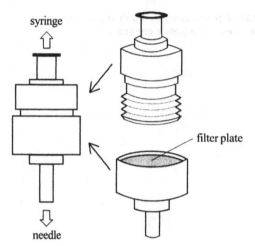

syringe

filter plate

needle

Fig. 1.21. Filtration assembly consisting of a syringe, a Swinny-type filter holder, and a syring needle

into the sample tube. A short glass tube with a sintered glass plate at one end can be very conveniently used for filtering the sample solution (Fig. 1.20). The filter plate must be moistened with a small amount of the solvent to make the filtration smooth. Filtration under nitrogen pressure can be easily done with this assembly in the case of very viscous solutions. Filtration under pressure can also be carried out using a rubber bulb with a pressure reservoir.

Another useful filter assembly is a small gas-tight syringe fitted with a Swinny-type filter holder and a thin syringe needle. The syringe and the filter holder are fitted with a Luerlock (Fig. 1.21). Teflon filters with 0.5-μm pores are usually used for NMR experiment; the filters will tolerate almost all the solvents used in NMR work.

Not only the concentration of the sample solution but also the volume placed in an NMR sample tube is very important to obtain a spectrum with high resolution. Too small a volume causes the field distortions by the interface between the solution and air, and degrades the line-shape and resolution, and consequently the sensitivity. It is necessary to find out the minimum volume or depth of the sample solution which gives adequate resolution on a particular NMR probe and to adjust the positional setting of the sample tube optimally in the spinner turbine. A sample volume larger than the minimum one is essentially noncritical; however, it is preferable for obtaining good spectral resolution easily to identify the optimum sample volume for a particular probe and to prepare samples always with a consistent depth of solution.

Sample tubes should usually be sealed by appropriate techniques. For ordinary measurements, commercial plastic caps are enough to prevent evaporation of the solvent if the temperature of measurement is lower than the boiling point of the solvent. Special caps for precision coaxial tubing (Fig. 1.18) are also commercially available. A sample tube sealed with a polyethylene shrinkable tube (Fig. 1.22) can be used at temperatures slightly higher than the boiling points of the solvents.

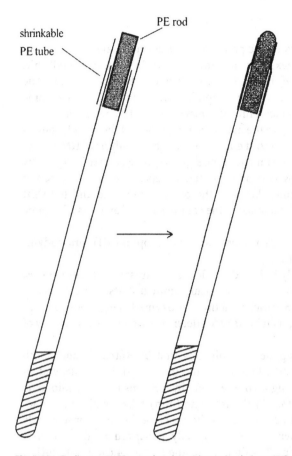

shrinkable
PE tube

PE rod

Fig. 1.22. Sealing an NMR sample tube with a polyethylene (*PE*) shrinkable tube

For special cases, such as measurements of spin–lattice relaxation times and of the spectra of air-sensitive compounds or those in which very high resolution is needed, the sample tube should be sealed using a small gas burner under vacuum or nitrogen pressure while cooling it in the bath. Special care must be taken so as to retain the cylindrical symmetry of the tube when it is sealed. In the case where the tube is sealed under nitrogen pressure, the tube should not be immersed too deeply in the liquid nitrogen bath. Deep immersion often causes condensation of a small amount of liquid nitrogen in the tube and this will explode when warmed up; the temperature of the liquid nitrogen in the Dewar vessel is sometimes decreased below the boiling point of nitrogen owing to the evaporation of liquid nitrogen.

1.4.4 NMR Shift Reagent

Since Hinckley [28] demonstrated the practical application of paramagnetic lanthanide β-diketoenolate complexes for inducing shifts in NMR spectra, NMR shift reagents have been found useful in the area of spectral clarification [29]. The lanthanide complexes associate with nucleophilic functional groups of organic compounds in solution and induce shifts of the peaks in NMR spectra of organic substrates predominantly owing to the magnetic interaction of the pseudocontact in nature. The lanthanide-induced shift enables a simplified analysis of NMR spectra with overlapping signals; assignment of the signals can be done more easily by using shift reagents. Since the pseudocontact shift data depend on the structure and geometry of the complex formed between the organic substrate and the shift reagent, the shift reagents can be applied to the conformational analysis of organic compounds in solution.

The most frequently used shift reagents are the europium(III), praseodymium(III), and ytterbium(III) chelates.

Europium reagents generally induce downfield shifts in substrate resonances, while the praseodymium analogues generally cause upfield shifts. Optically active shift reagents such as tris[(3-trifluoromethylhydroxymethylene)-(+)-camphorato]europium(III) [Eu(tfmc)$_3$] can be used for determination of optical purity of chiral organic compounds.

NMR shift reagents were applied to polymers first by Katritzky and Smith [30] and Guillet et al. [31] It was found that the induced shift was dependent on the tacticity of PMMA; the signal of the α-methyl protons in the syndiotactic (rr) triad shifted more than that in the isotactic (mm) triad, while the isotactic methoxy proton peak shifted more than the other peaks, and consequently the tacticity-sensitive peak separation was greatly enhanced in the presence of tris[1,1,1,2,2,3,3-heptafluoro-7,7-dimethyloctanedionato(4,6)]europium(III) [Eu(fod)$_3$]. The phenomena were confirmed and investigated in more detail by Amiya et al. [32].

Successful determination of triad tacticity by 100 MHz ^{1}H NMR spectra with Eu(fod)$_3$ was reported on cationically prepared poly(methyl vinyl ether) (PMVE) [33]. Addition of Eu(fod)$_3$ shifted all the peaks downfield and the extent of the shifts was in the order $CH > CH_2 > OCH_3$. The larger downfield shift for the methine signals than for the methoxy signal allowed the splitting of the methoxy signal due to the triad tacticity to be observed without interference from the methine proton multiplet. Fine splitting was clearly observed in the tactic triad peaks of the methoxy resonance and was ascribed to the pentad tacticity.

The measurements of the spectra were redone using a 500 MHz spectrometer. The results are shown in Fig. 1.23. The resolution of the spectra is greatly enhanced by use of a 500 MHz instrument compared with the case at 100 MHz. The downfield shifts increase linearly with increasing ratio of Eu(fod)$_3$ to PMVE, and splitting due to the pentad tacticity was observed at ratios of 0.02 and 0.04.

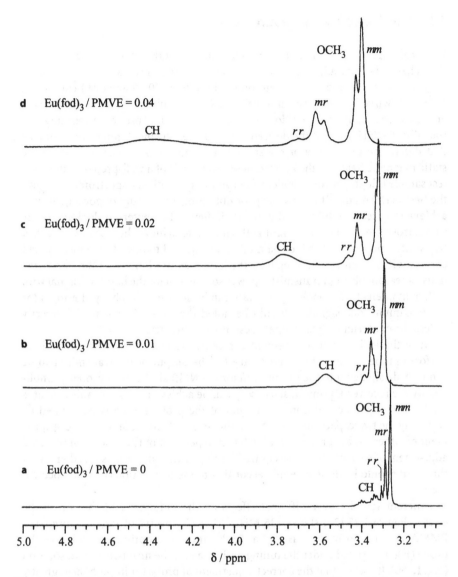

Fig. 1.23a–d. 500 MHz ¹H NMR spectra of PMVE in carbon tetrachloride containing 10% benzene-d_6 measured in the presence of Eu(fod)$_3$ at 60 °C [25]. 45° pulse, pulse repetition time 10 s, 16–64 scans

Generally, the peak width increases as the ratio increases. At higher temperature the peaks become sharper, while the induced shifts slightly decrease. So it is very important to optimize the effects of these different factors. The tacticity dependence of the induced shift (syndiotactic > heterotactic > isotactic) may be due to the difference in the conformation of the tactic sequences.

1.5   Outline of NMR Measurement

Practical implementation of NMR measurement is outlined in Fig. 1.24 [25] as a
flow chart.[3] First of all, a deuterated solvent solution of a standard test sample
[e.g., a deuterated chloroform solution of chloroform (0.26 wt/vol%) containing
TMS (0.22 wt/vol%)] degassed and sealed in an NMR tube is set in the NMR probe,
and the tube is spun by an air-flow turbine system at an appropriate spinning rate
(usually 15–20 Hz) and at the desired temperature. The most important elements
that determine the quality of the spectrum are the sample itself as well as the
static magnetic field and the spectrometer. So the use of a well-prepared standard
test sample is highly recommended when you start to adjust a spectrometer to give
the best performance. It is necessary for obtaining a spectrum of good quality to
achieve long-term stability of the magnetic field. The deuterium lock using the
signal from the deuterated solvent is the means to achieve the stability. The lock
transmitter power and the lock gain must be adjusted to operate the lock system
correctly. It is essential not to operate the lock at saturation and at the same time
to use a reasonably high transmitter power so as to obtain the best lock signal with
sufficient intensity. The lock signal phase can be adjusted simply by altering it for
the maximum lock signal. It should be noted that this will not work correctly
unless the deuterium signal shape is reasonably Lorentzian.

Once the lock system is operated correctly, the next thing to be done is to
perform probe tuning. This is necessary for the output of the transmitter to be
properly dissipated into the sample and for the NMR signal to be properly amplified by the receiver. Optimum sensitivity can be achieved when the adjustment is
made correctly. Practical implementation of the probe tuning is performed by
adjusting the two capacitors mounted in the resonant circuit of the probe. Adjustment of either capacitor is not completely independent of the other, and repeated
adjustment of each capacitor is required for the optimization. This is rather tricky
thing but should be done carefully according to the instructions of your spectrometer.

Different samples have different effects on the inductance of the coil in the
circuit. Figure 1.25 [25] shows 125 MHz ^{13}C NMR spectra of radically prepared
PMMA in chloroform-d measured at 55 °C under the optimized tuning of the
probe (Fig. 1.25a) and under the tuning optimized for the nitrobenzene-d_5 solution
(Fig. 1.25b). It is clear that the correct adjustment of probe tuning enhances greatly
the S/N ratio of the spectrum. So careful adjustment of the probe tuning is very
important for obtaining a spectrum of good quality, particularly in the measurement of a sample of low concentration or that of the nucleus of low natural abundance [25].

The lock system, once correctly optimized, provides information about the
homogeneity of the magnetic field. High homogeneity can be achieved by adjust-

---

[3] For further advanced discussion on instrumental and experimental details of NMR
measurement, refer to Ref. [5].

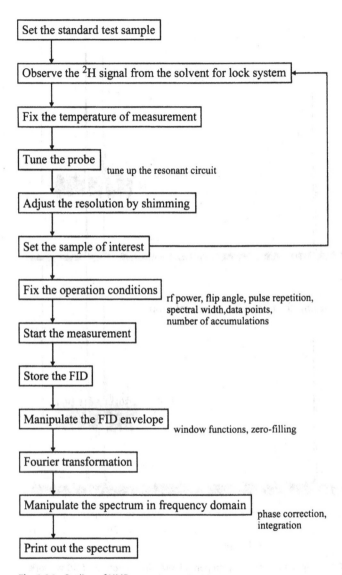

Fig. 1.24. Outline of NMR measurement

ing the intensity of the lock signal to the maximum. The adjustment can be made by altering the shim coil current (shimming). Ordinary NMR users may be concerned with Z shims (e.g., Z^1, Z^2, Z^3, etc.) and X and Y shims (e.g., X, Y, XY etc.) as far as the spectrometer is well maintained. The field of the Z shims coil is aligned along the vertical axis of the magnet, while that of X and Y shims is aligned along the other two orthogonal axes. Resolution is the fundamental aspect of spectrometer performance and is specified by the half-height width of a particular line measured in hertz and the extent to which the lineshape deviates from an ideal

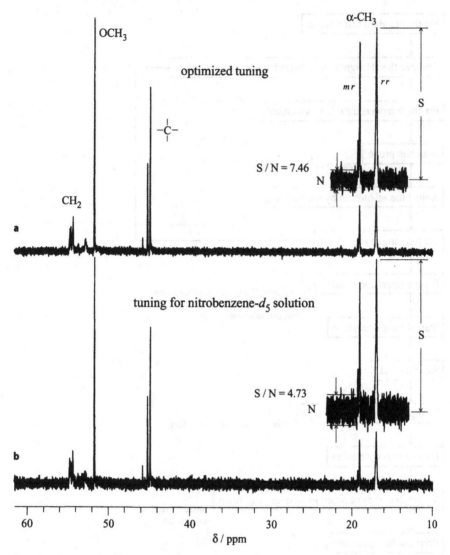

Fig. 1.25a,b. 125 MHz ^{13}C NMR spectra of radically prepared PMMA in chloroform-*d* (10 wt/vol%) measured at 55 °C. Effect of adjusting the probe tuning [25]. 45° pulse, pulse repetition time 3 s, 600 scans

Lorentzian. The latter can be examined by the linewidth at the height of the ^{13}C satellite of the ^{1}H signal of chloroform (0.55% of the peak amplitude) and one-fifth this height (0.11% of the peak amplitude). The expected linewidth at 0.55% is 13.5 times the half-height width and that at 0.11% 30 times (Fig. 1.26).

The linewidth and the lineshape are affected strongly by Z^1, Z^2, Z^3, and Z^4 and sometimes Z^5. Mis-adjustment of Z^1, Z^3, and Z^5 causes symmetrical broadening of the peak and that of Z^2 and Z^4 causes unsymmetrical broadening. The difficult

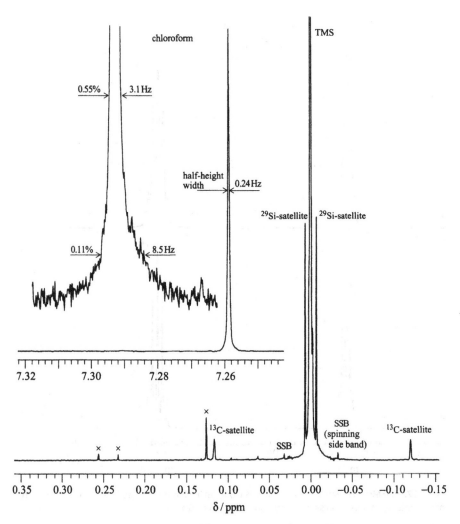

Fig. 1.26. 500 MHz ^{1}H NMR spectra of chloroform in chloroform-*d* (chloroform 0.26 wt/vol%, TMS 0.22 wt/vol%). 45° pulse, pulse repetition time 15 s, 32 scans, 25 °C

thing in the shimming or adjustment of shim coils often arises from the existence of interaction between certain combinations of the shims. So the essential part of the skill of shimming is to know how to treat such necessary interactions. It is necessary to adjust the combination of Z^1 and Z^2 all the time. It should be noted that the outcome of adjusting the higher-order shim, Z^2, is not meaningful alone. It is essential to reoptimize the lower component, Z^1, and then judge whether or not the combined effect is an improvement. These adjustments are bothersome and time-consuming but very important for the best performance of the spectrometer and are now rather easily made with the aid of a computer.

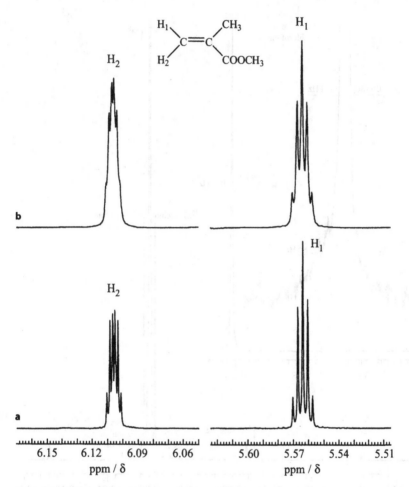

Fig. 1.27. 500 MHz ^{1}H NMR spectra of MMA (5 wt/vol%) in chloroform-d at 25 °C **a** with and **b** without sample spinning. 45° pulse, pulse repetition time 15 s, 16 scans

The intensities of the spinning side bands can be minimized by adjusting X and Y shims; the first-order side bands by X, Y, XZ, and YZ shims and the second order sidebands by XY and X^2-Y^2 shims.

The 500 MHz ^{1}H NMR spectra of the standard test sample containing 0.26 wt/vol% chloroform and 0.22 wt/vol% TMS in chloroform-d measured at 35 °C after the adjustment is shown in Fig. 1.26. The following are the recommended criteria; the half-height width of chloroform should be less than 0.5 Hz, the peak heights of ^{13}C or ^{29}Si satellites or the spinning sidebands, each appearing on both sides of the corresponding main signals, should be almost the same, the peak heights of the spinning sidebands should be smaller than those of the corresponding ^{13}C satellites. The peak widths of the chloroform signal at the 0.55 and 0.11% points are 3.1 and 8.5 Hz, respectively, and so the expected half-height width for a Lorentzian

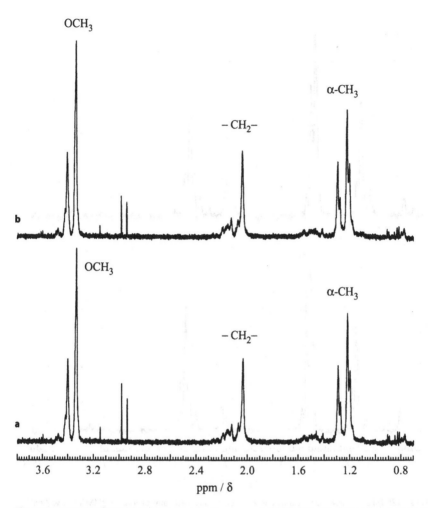

Fig. 1.28. 500 MHz ¹H NMR spectra of radically prepared PMMA (10 wt/vol%) in benzene at 45 °C **a** with and **b** without sample spinning. 45° pulse, pulse repetition time 5 s, 512 scans

is 0.23 and 0.28 Hz, which is close to the observed value of 0.24 Hz. All these criteria are fulfilled in the spectra shown in Fig. 1.26.

When the adjustment for the standard test sample is completed, the sample of interest is introduced in the probe at the required temperature of measurement and the same procedure of adjustment is repeated. If the solvent is the same as that for the test sample, the differences are relatively minor and the readjustments of Z^1 and Z^2 shims are enough to achieve the high performance. In the case where the character of the solvent is very different from that for the test sample, the probe tuning and the more extensive readjustment of the shims are required. It should be noted that the sample depth affects strongly the required shim values. So it is highly recommended to prepare samples with a consistent depth of solution (Sect. 1.4.3).

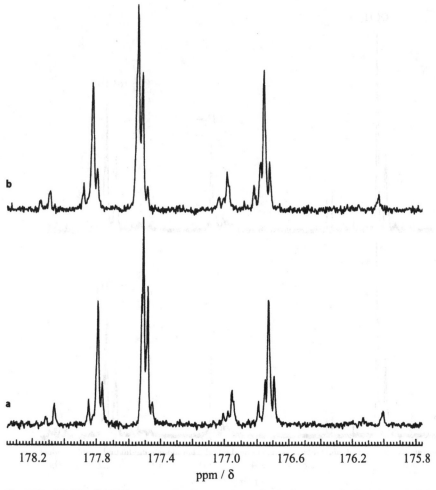

Fig. 1.29. 125 MHz ^{13}C NMR spectra of carbonyl carbon of radically prepared PMMA (10 wt/vol%) in toluene-d_8 at 90 °C **a** with and **b** without sample spinning. 45° pulse, pulse repetition time 5 s, 12,000 scans

After readjusting the probe for the sample of interest, the following operation conditions should be set up before starting the accumulation of the FIDs; RF power (flip angle), pulse repetition, spectral width, data points, and number of accumulations. The effects of these elements of the operating conditions on the spectrum are described in Sect. 2.1. When the accumulation of the transients is over, the accumulated FID signal is stored in computer memories. The decay envelope of the FID is then manipulated by using window functions (Sect. 2.1) to balance or optimize the sensitivity and resolution of the spectrum transformed into the frequency domain. Application of the best window function is essential to reach the utmost limit of the spectrometer. Several functions, such as zero-filling, matched filter, and Lorentz–Gauss transformation, have been designed and applied to improve sensitivity or resolution. The manipulated FID is then transformed into the frequency-domain spectrum and it is printed out after phase correction, with integration of spectra (information of peak area) if desired. The FID data have to be stored in case other information is needed on the spectrum.

As described in Sect. 1.2, the sample tube is rotated using an air-flow turbine about its axis to average out the magnetic field observed at each part of the sample, producing increased resolution of the spectrum. The enhancement of resolution by spinning is remarkable in the case of a low molecular weight compound as shown in Fig. 1.27. However, in the case of a polymeric compound with a modern high-performance NMR spectrometer, the ^{1}H NMR spectral resolution is quite good; the decrease in the linewidth by spinning is only small (Fig. 1.28). In the measurement of the ^{13}C NMR spectrum of a polymer the resolution enhancement by spinning is observed for several peaks (e.g., 117.5 ppm) as shown in Fig. 1.29.

References

1. EMSLEY JW, FEENEY J, SUTCLIFFE LH (1965) High resolution nuclear magnetic resonance spectroscopy. Pergamon, Oxford
2. BECKER ED (1980) High Resolution NMR: theory and chemical applications, 2nd edn. Academic, New York
3. YODER CH, SCHAEFFER CD JR (1987) Introduction to multinuclear NMR: theory and application. Benjamin/Cummings, Menlo Park, CA
4. HARRIS RK (1983) Nuclear magnetic resonance spectroscopy: a physicochemical view. Pitman, Marshfield, MA
5. DEROME AE (1987) Modern NMR techniques for chemistry research. Pergamon, Oxford
6. (a) GRANT DM, HARRIS RK (eds) (1996) Encyclopedia of NMR, vols 1–8. Wiley, New York; (b) GRANT DM, HARRIS RK (eds) (2002) Encyclopedia of NMR, vol 9. Advances in NMR. Wiley, New York
7. WEBB GA (1968–2002) Annual reports on NMR spectroscopy, vols 1–45. Academic, London
8. WEBB GA (ed) (1972–2002), Nuclear magnetic resonance, vols 1–31. The Royal Society of Chemistry, London
9. NOGGLE JH, SCHIRMER RE (1971) The nuclear Overhauser effect. Chemical applications. Academic, New York
10. ANDO I, ASAKURA T (1998) Solid state NMR of polymers. Elsevier, Amsterdam
11. RANDALL JC (1977) Polymer sequence determination – Carbon-13 NMR method. Academic, London
12. WÜTHRICH K (1975) NMR in biological research: peptides and proteins. North-Holland, Amsterdam, p 334
13. DANIELSON A, FALKE J (1996) Annu Rev Biophys Biomol Struct 25:163
14. BEVINGTON JC, HUCKERBY TN, VICKERSTAFF N (1983) Makromol Chem Rapid Commun 4:349
15. TONELL AE (1989) NMR spectroscopy and polymer microstructure – The conformational connection. VCH, Weinheim, p 97
16. OVENALL DW, FERGUSON DW (1987) In: Breyed WS (ed) Pulse methods in 1D and 2D liquid phase NMR. Academic, New York, chap 7

17. GORENSTEIN DG (1984) Phosphorus-31 NMR, principles and applications. Academic, New York
18. KOBAYASHI S, SUZUKI M, SAEGUSA T (1984) Macromolecules 17:107
19. LEVY GC, LICHTER RL (1979) Nitrogen-15 nuclear magnetic resonance spectroscopy. Wiley, New York
20. WITANOWSKI M, STEFANIAK L, WEBB GA (1986) Annu Rep NMR Spectrosc 18
21. HATADA K, UTE K, KASHIYAMA M (1990) Polym J 22:853
22. HATADA K, TERAWAKI Y, KITAYAMA T (1992) Kobunshi Ronbunshu 49:335
23. HATADA K, TERAWAKI Y, OKUDA H, NAGATA K, YUKI H (1969) Anal Chem 41:1518
24. HATADA K, TERAWAKI Y, OKUDA H (1977) Org Magn Reson 9:518
25. HATADA K, KITAYAMA T, TERAWAKI Y, NISHIURA T (1995) Kobunshi Jikkengaku 5:23
26. FLATH RA, MENDERSON N, LUNDIN RE, TERANISHI R (1967) Appl Spectrosc 21:183
27. HATADA K, TERAWAKI Y, OKUDA H (1969) Bull Chem Soc Jpn 42:1781
28. HINCKLEY CC (1969) J Am Chem Soc 91:5160
29. SIEVERS RE (ed) (1973) Nuclear magnetic resonance shift reagent. Academic, New York
30. KATRITZKY AR, SMITH A (1971) Tetrahedron Lett 1765
31. GUILLET JE, PEAT IR, REYNOLDS WF (1971) Tetrahedron Lett 3493
32. AMIYA S, ANDO I, CHUJO R (1973) Polym J 4:385
33. YUKI H, HATADA K, HASEGAWA T, TERAWAKI Y, OKUDA H (1972) Polym J 3:645

2 Measurement of Spectrum with High-Quality Including Quantitative Analysis

2.1 Effects of Operating Conditions

NMR is uniquely powerful in the determination of polymer microstructure, such as stereochemical configuration (tacticity), geometrical isomerism, regioirregularity, and monomer sequences in copolymers. Quantitative data concerning these structural characteristics of polymers can be obtained from relative intensities of the respective peaks for these structures, and are hardly collected by other analytical means. NMR measurements for the analyses of polymers should be performed with much higher accuracy and precision than those for low molecular weight compounds; in the latter only approximate intensity ratios, such as $CH_3:CH_2=3:2$, are required.[1] Experimental conditions for quantitative polymer analysis have been discussed in reviews [1, 2] and those for general purposes in a book by Derome [3].

Hatada and coworkers [4–7] have studied the precision and the accuracy of NMR studies of polymers using CW spectrometers. The effects of the measurement conditions, including temperature, sample concentration, and the RF field strength, were examined using several polymer and copolymer samples [5]. It was found that the optimum conditions depended on the samples and that under the optimized conditions the errors in intensity measurements were less than 5%.

Nowadays almost all NMR data are acquired by the pulsed FT method. In this method, the consideration of the spin–lattice relaxation times (T_1) of the signals of interest, i.e., proper selection of the flip angle (pulse width) of the observed pulse and the pulse repetition time is of primary importance for quantitative analysis [8–11].

---

[1] Accuracy deals with the deviation from the theoretical value and precision with the fluctuation of the measurements. We regard the results as accurate if the mean value of the determinations agrees well with the theoretical value, while we regard the results as precise if the results of determinations agree well mutually. In this book accuracy is expressed by $100 \times (|x_t - \bar{x}|)/x_t$ (%) and precision by $100 \times \sigma/\bar{x}$ (%), where x_t, x_i, $\bar{x}$, n, and σ represent the theoretical value, the value of an individual determination, the mean of the individual results, and the number of determinations, the standard deviation $\sigma [= \sqrt{\sum (x_i - \bar{x})^2/n}]$, respectively.

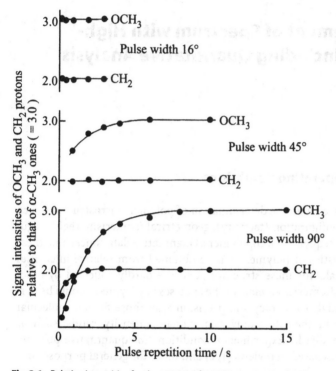

Fig. 2.1. Relative intensities for the 500 MHz ^{1}H NMR signals of radically prepared PMMA against pulse repetition times at several flip angles [2]. 10 wt/vol% CDCl$_3$ solution, 55 °C under nitrogen, 16 scans. Signal intensities are represented relative to that for α-CH$_3$ proton signals (=3.00). T_1 for OCH$_3$, CH$_2$, and α-CH$_3$ protons under the conditions of the measurement are 1.3, 0.35–0.40, and 0.25–0.26 s, respectively

Effects of flip angle and pulse repetition time on relative peak intensities in the ^{1}H NMR spectrum were examined for radically prepared PMMA (Eq. 2.1) in chloroform-d (10 wt/vol%) under nitrogen pressure at 55 °C, using a 500 MHz NMR spectrometer [12].

$$-\left(CH_2-\underset{\underset{\underset{OCH_3}{|}}{\overset{\overset{CH_3}{|}}{\underset{C=O}{|}}}{C}\right)_n \qquad (2.1)$$

The relative intensities for methylene and methoxy protons against that of α-methyl protons with several repetition times and flip angles are shown in Fig. 2.1. At a flip angle of 16° the relative intensities of the OCH$_3$ (T_1=1.3 s) and CH$_2$ (T_1=0.35–0.40 s) proton signals against α-CH$_3$ (T_1=0.25–0.26 s) proton signals are close to the theoretical values even at a pulse repetition time of 0.2 s. With a flip angle of 90°, the intensity of the OCH$_3$ signal is 97% of the theoretical value

Table 2.1. Mean values of relative intensities of ¹H NMR and ¹³C NMR signals of PMMA in CDCl₃ at 55 °C for five runs [10 wt/vol% of solution, values in parentheses are standard deviations σ (%)] [12]

¹H(α-CH₃=3.00)[a]		¹³C(α-CH₃=1.00, COM)[b,c]				¹³C(α-CH₃=1.00, NNE)[b,d]				¹³C(α-CH₃=1.00, NON)[e]			
OCH₃	CH₂	C=O	CH₂	OCH₃	Quat.C	C=O	CH₂	OCH₃	Quat.C	C=O	CH₂	OCH₃	Quat.C
3.022	2.002	0.482	0.772	0.787	0.812	0.978	1.018	1.082	1.087	1.003	1.020	1.037	1.030
(0.8)	(1.8)	(1.9)	(2.0)	(2.4)	(2.6)	(1.5)	(1.5)	(2.1)	(1.8)	(2.2)	(2.1)	(1.5)	(1.0)

[a] Frequency 500 MHz, pulse width 45°, repetition time 10 s, 4–16 scans.
[b] Frequency 125 MHz, pulse width 45°, repetition time 5 s, 5,000 scans.
[c] Measurement under a complete decoupling with NOE (COM).
[d] Measurement under a gated decoupling without NOE (NNE).
[e] Measurement without ¹³C–¹H decoupling (NON).

when the pulse repetition time is 6 s (about 5 times T_1, see Sect. 1.2) and reaches the theoretical value with a pulse repetition time as long as 10 s. With a flip angle of 45° and a pulse repetition time longer than 6 s, the intensities of all the signals were almost 100% of the theoretical values. Both the accuracy and the precision of the measurement with the flip angle of 45° are quite high as shown in Table 2.1. The results indicate that measurement at a flip angle as small as possible is preferable for the accurate quantitative analysis if the S/N ratio of the peaks is large enough.

In the case of low molecular weight compounds the T_1 of the protons are longer than those for polymers and a much longer pulse repetition time is required. For example, the ^{1}H T_1 for C_6H_5, CH_2, and CH_3 groups of ethylbenzene are 32.3, 15.5, and 12.9 s (2% CDCl$_3$ solution, at 35 °C and 500 MHz, under nitrogen), respectively, and pulse repetition times of 60–70 and 120–150 s were necessary for accurate measurement of the signal intensity with flip angles of 45° and 90°, respectively (Fig. 2.2). The results show that a repetition time of 5 times the longest T_1 is necessary for the accurate quantitative measurement in the case of the 90° pulse (Sect. 1.2) but a repetition time of about 2 times the longest T_1 is sufficient in the case of the 45° pulse. The difference of the results for the 45° pulse between PMMA

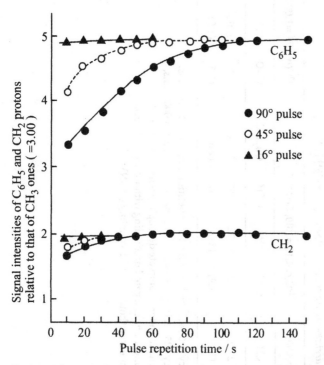

Fig. 2.2. Relative intensities for the 500 MHz ^{1}H NMR signals of ethylbenzene against pulse repetition times at several flip angles. 2 wt/vol% CDCl$_3$ solution, 35 °C under nitrogen, four scans. Signal intensities are represented relative to that for CH$_3$ proton signals (=3.00). T_1 for C$_6$H$_5$, CH$_2$, and CH$_3$ protons are 32.3, 15.5, and 12.9 s, respectively

Table 2.2. Measurement of ^1H NMR signal intensities of $(CH_3)_3Al$ (neat) and $CHCl_3$ in benzene-d_6 (80 vol/vol%) placed in the inner and outer cells of coaxial tubing, respectively, at various pulse repetition times (55 °C, 100 MHz, $^1H T_1$ for CH_3 and CH protons are 4.77 and 4.57 s, respectively) [2]

Pulse repetition time (s)	Relative intensities (CH_3/CH)[a]
5	1.054
10	1.059
15	1.061
20	1.062
30	1.065
60	1.056

[a] Calculated value for the intensity ratio of CH_3 and CH signals is 1.053.

and ethylbenzene may be due to the ratio of the T_1 of interest to the longest T_1. The closer the ratio to unity, the shorter the repetition time necessary for accurate intensity measurement. A detailed explanation with an example is given in the following paragraph. For this ethylbenzene solution a single-pulse measurement (1 scan) with a 45° or a 90° pulse provided a spectrum with a large enough S/N ratio, which gave a quantitative result of high accuracy and precision of 0.5–1.5%. At the flip angle of 16°, accurate measurements could be done even with repetition times of 10–20 s.

In the case of long T_1 for two signals of interest but with a small difference between the T_1, or exactly speaking, with the ratio between the T_1 close to unity, a short repetition time is enough to obtain the relative intensity with high accuracy. A typical example is shown in Table 2.2. In this example, the intensities of the signals due to the methyl protons of trimethylaluminum (neat) and the methine proton of chloroform (80 vol/vol% C_6D_6 solution), which were placed in the inner and outer cells of coaxial tubing (Fig. 1.18b), respectively, were determined. The T_1 values for the former and the latter signals were 4.77 and 4.57 s, respectively, but the repetition time of 5 s is enough to obtain the theoretical value for the relative intensity.

Measurements of NMR spectra under air greatly decrease T_1, particularly in the case of low molecular weight compounds. For example, the T_1 of C_6H_5, CH_2, and CH_3 protons of ethylbenzene for the previously mentioned solution (2% $CDCl_3$ solution, 35 °C, 500 MHz) in air were 7.42, 6.15, and 5.49 s, respectively. The decrease in T_1 decreases the minimum repetition time required for the accurate quantitative determination and a repetition time of 30–40 s is enough for the accurate measurement with the flip angle of 90°. Conversely, the repetition time appropriate for the accurate measurement under air is usually too short for the measurement under nitrogen or in vacuo. A typical example is shown in Table 2.3 for ^1H NMR measurement of methyl acrylate. This point is sometimes overlooked, leading to erroneous determination of signal intensities.

Table 2.3. Intensity measurements in the 500 MHz ^{1}H NMR spectrum of methyl acrylate in CDCl$_3$(5.0 wt/vol%) in nitrogen and in air at 30 °C

Atmosphere	Repetition time (s)[a]	Signal intensities[b]		
		H$_1$	H$_2$	H$_3$
N$_2$	10	0.95	0.92	0.81
N$_2$	30	1.00	0.98	0.94
N$_2$	60	1.00	1.00	0.99
Air	10	1.00	0.99	0.96

[a] T_1 in nitrogen: H$_1$ 9.4 s, H$_2$ 11.1 s, H$_3$ 20.25 s, OCH$_3$ 7.9 s. T_1 in air: H$_1$ 4.8 s, H$_2$ 5.2 s, H$_3$ 6.5 s, OCH$_3$ 4.7 s.
[b] Referred to the intensity of the OCH$_3$ signal (3.00).

In ^{13}C NMR measurement, consideration of the NOE for each carbon signal is also required in the quantitative determination (see Sects. 1.1.2, 7.1). The relative intensities of the ^{13}C NMR signals of PMMA were measured under various conditions using the same spectrometer used for the ^{1}H NMR measurements mentioned earlier and it was operated at a frequency of 125 MHz at 25 °C [12]. With a flip angle of 45°, the relative intensities of C=O (T_1=1.12 s), CH$_2$ (T_1=0.175 s), OCH$_3$ (T_1=1.12 s), and quaternary carbon (T_1=2.25 s) signals against the α-CH$_3$ signal (T_1=0.15 s) under the complete ^{1}H-decoupling condition (COM) reached constant values with a pulse repetition time of 5 s. The values of the intensities relative to the α-CH$_3$ intensity were less than unity (theoretical value) owing to the larger NOE value for α-CH$_3$ than for other carbons. The relative intensities observed with a gated decoupling mode without NOE (NNE) are close to unity with deviations of 1.8–8.7% from the theoretical value. Standard deviations for five runs are in the range 1.5–2.1% (Table 2.1). It is noteworthy that the NOE values for each carbon signal at 22.5 and 25 MHz are close to theoretical maximum, so the relative intensities observed under the COM (broadband decoupling) are close to the theoretical value [13] (see Table 2.5).

^{13}C NMR spectra are usually obtained with the accumulation of many transients owing to the low sensitivity and low natural abundance of ^{13}C nuclei. It should be noted that, as a practical matter, the S/N ratio of interest is not the S/N ratio of the signals obtained by a single scan but that obtained with accumulations during a certain time. In this sense, it is desirable to take the spectrum with as many accumulations as possible unless signal saturation (Sect. 7.1) does not occur. As described in Sect. 7.4 the values of the ^{1}H- and ^{13}C T_1 increase with an increase in resonance frequencies. This sometimes causes a decrease in the possible number of accumulations during a certain time in the measurement at high magnetic field

Table 2.4. Relation between the S/N ratios and pulse repetition times in the ^{13}C NMR spectra of radically prepared PMMA measured during a certain period of time (30 min) at 55 °C and different frequencies (10 wt/vol% CDCl$_3$ solution)

Pulse repetition time (s)	25 MHz			67.5 MHz			100 MHz		
	C=O	OCH$_3$	CDCl$_3$	C=O	OCH$_3$	CDCl$_3$	C=O	OCH$_3$	CDCl$_3$
3	2.3	5.1	0.3	7.2	26.4	2.8	7.5	31.3	4.6
5	2.1	3.8	0.7	5.8	19.5	2.9	6.5	27.7	5.7
10	1.3	2.9	1.1	4.3	13.4	3.7	4.3	17.9	7.5
20	0.7	2.1	1.3	3.2	11.1	5.9	2.8	12.3	9.9
30	1.2	1.5	1.5	2.8	8.5	6.7	2.4	10.0	10.0
40		1.6		2.2	7.5	7.5	2.1	9.3	12.7
50				2.0	6.3	7.5	1.9	8.1	13.6
60				2.1	6.0	8.5	1.9	7.5	14.4
120				1.3	4.2	9.7	1.3	5.2	15.5
180				1.8	3.7	10.2			
360				1.1	2.4	8.0			

and cancels the favorable effect of high magnetic field on the S/N ratio. The S/N ratios of ^{13}C NMR signals of PMMA measured in CDCl$_3$ at 55 °C and 125 MHz for 30 min with 90° pulses and different repetition times, i.e., different numbers of accumulations, are illustrated in Fig. 2.3. The results indicate that the S/N ratios increase with decreasing repetition time down to about 5 s, i.e., close to twice the T_1 of the quaternary carbon, which is the largest among the ^{13}C T_1 for PMMA in CDCl$_3$. So if a spectrum with a larger S/N ratio but not with necessarily quantitative data is required, a repetition time as short as possible should be used. Similar results were obtained in the measurements at 25, 67.5, and 100 MHz (Table 2.4). The S/N ratio for the CDCl$_3$ signal at 125 MHz, whose ^{13}C T_1 is much longer (117 s) than that for PMMA, increases with decreasing repetition time down to about 250 s and then decreases. The S/N ratio for quaternary carbon (^{13}C T_1=2.25 s) turns into a decrease at a repetition time of 2 s (Fig. 2.3). The optimum repetition time for the carbon of CDCl$_3$ or the quaternary carbon of PMMA decreased with decreasing resonance frequency owing to the decreases in ^{13}C T_1, and accordingly the available number of accumulations increased.

The S/N ratios increase with an increase in the flip angle up to 90° for all the carbons of PMMA in the measurements with a repetition time of 3 s for 30 min (600 accumulations) at 55 °C and 125 MHz (Fig. 2.4). The S/N ratio for CDCl$_3$ having much longer ^{13}C T_1 (117 s) first increases with increasing flip angle and reaches a maximum at a flip angle of about 20°, and then decreases owing to saturation as shown in the figure. On the other hand, the S/N ratios of PMMA carbon signals increase with increasing flip angles for all the carbons. So in the measurement of the ^{13}C NMR spectrum of the polymer the flip angle of 90° is recommended to obtain the signals with high S/N ratios. Particularly, in the ^{13}C NMR spectrum measured

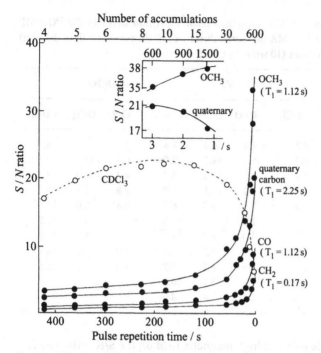

Fig. 2.3. S/N ratios of the signals against pulse repetition times and number of accumulations for 125 MHz ^{13}C NMR spectra of radically prepared PMMA measured with a 90° pulse during a certain period of time (30 min) [2]. 10 wt/vol% CDCl$_3$ solution, 55 °C

under the COM at higher magnetic fields, for example, at 125 MHz, the signal intensities usually deviate from the theoretical value owing to the different NOE values for each carbon. In such a case quantitative determinations are not of interest and thus measurements with the flip angle of 90° and the repetition time as short as possible are highly recommended.

The S/N ratio for the ^{13}C NMR spectrum of PMMA was studied at different frequencies with a flip angle of 90° and a repetition time of 3 s for 30 min. The results are shown in Fig. 2.5. The S/N ratios increase remarkably with increasing resonance frequency up to 67.5 MHz and then increase slightly. The large difference in the S/N ratio observed at 25 and 67.5 MHz is partly due to the difference in the effective detection volume for the sample solution. The S/N ratio obtained at 100 MHz is close to that obtained at 125 MHz. This means that a 100 MHz spectrometer (400 MHz for ^{1}H) is practically very useful at present if the peak separation is enough for the analysis.

The NMR signal in FID experiments decays with time, while the noise amplitude remains constant. As a result, suppressing the relative contribution of the tail end of the FID improves the S/N ratio of the Fourier-transformed spectrum. This can be achieved by multiplying the FID data by a decreasing exponential function (Eq. 2.2, BF>0), where t is the time and BF is the broadening factor (Hz):

$$E(t) = \exp[-(BF)\pi t]. \tag{2.2}$$

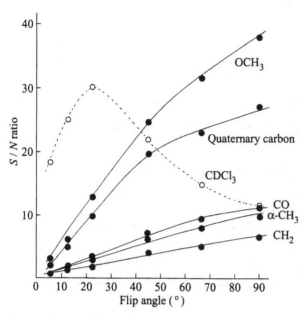

Fig. 2.4. S/N ratios of the signals against flip angles for 125 MHz ^{13}C NMR spectra of radically prepared PMMA measured during a certain period of time (30 min) [2]. 10 wt/vol% CDCl$_3$ solution, 55 °C, pulse repetition time 3 s, 600 scans

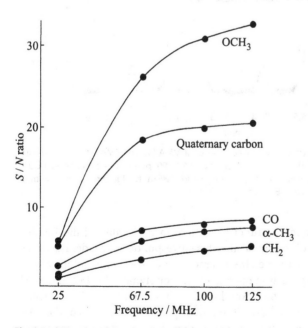

Fig. 2.5. S/N ratios of the signals for ^{13}C NMR spectra of radically prepared PMMA measured at different frequencies [2]. 10 wt/vol% CDCl$_3$ solution, 55 °C, 90° pulse, pulse repetition time 3 s, 600 scans

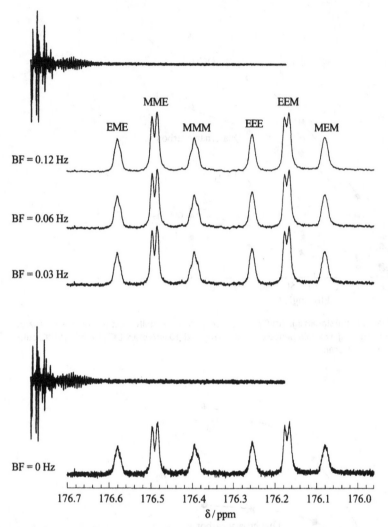

Fig. 2.6. 125 MHz ^{13}C NMR spectra of isotactic copolymer of MMA (*M*) and EMA (*E*) measured with different broadening factors [2]. 10 wt/vol% CDCl$_3$ solution, 55 °C, 90° pulse, pulse repetition time 3 s, 2,400 scans, digital resolution 0.03 Hz, Codes such as *EME* and *MME* in the figure represent the triad sequences of monomeric units (M and E) (Sect. 4.1, Table 4.3)

Multiplication by the exponential function with a positive value of the BF also speeds the apparent decay of the signal and this corresponds to broadening of the signal in the frequency domain, i.e., the degradation of spectral resolution. The NMR signals of the carbonyl carbon of an isotactic copolymer of MMA and ethyl methacrylate (EMA) measured with several BFs are shown in Fig. 2.6. The attached FIDs are those with (BF=0.12 Hz) and without multiplication; one tenth of the FIDs from the beginning are shown. It is very clear that the S/N ratio increases with increasing value of the BF in association with decreasing resolution of the signals.

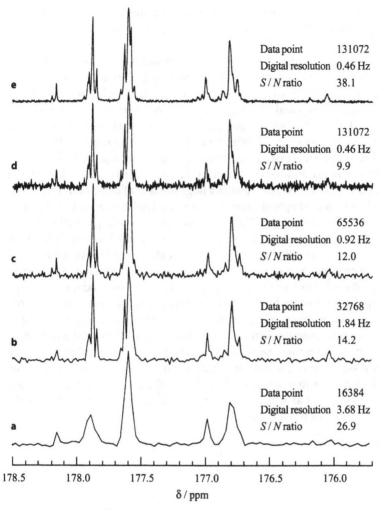

Fig. 2.7a–e. 125 MHz ^{13}C NMR spectra of radically prepared PMMA measured with different data points [2].10 wt/vol% CDCl$_3$ solution, 55 °C, 90° pulse, pulse repetition time 3 s, spectral width 30.146 KHz, 600 scans (**a–d**), 9,600 scans (**e**). The S/N ratios are those of the peak at 177.6 ppm

The optimum balance appears to be obtained when the spectrum is manipulated with a BF of 0.06 Hz. It is important to choose the optimum value of BF to obtain a spectrum of good quality. Application of a negative value of BF suppresses the decay of the FID and provides us with a resolution enhancement of the spectrum. The resulting amplification of the tail of the FID increases the noise level and large wiggles due to truncation of the FID are likely to arise. The resolution of these problems can be achieved by using a function that cancels the decay of the early part of the FID and falls smoothly towards zero by the end, in combination with an exponential function. Several combinations of the functions are used for measurement with resolution enhancement. It should be noted that the resolution enhancement sometimes causes a decrease in the accuracy of the quantitative determination.

In order to observe the fine structure of the spectrum, it is necessary to improve the digital resolution. This is achieved by increasing the number of data points or by reducing the spectral width as seen from Eq. 1.12. The 125 MHz ^{13}C NMR spectra of radically prepared PMMA were measured with a spectral width of 241.17 ppm and with different data points. The carbonyl carbon resonance range of the spectra is shown in Fig. 2.7. With an increase in the number of data points the spectral resolution increases in association with an increase in the noise level (Fig. 2.7a–d) and a large number of accumulations are needed to obtain a spectrum of high S/N ratio in the measurement with a large number of data points (Fig. 2.7e). This is also the problem of optimization. The enhanced noise level is due to the long acquisition time for sampling the FID signals acquired with a large number of data points. The noise level in the FID is constant, while the real NMR signals from the sample decay with time. Thus the noise components in the FID signal increase as the acquisition time increases, and the S/N ratios of the signal in the frequency domain are lowered. Most of the NMR signals from polymer samples undergo relatively rapid relaxation and the FID signals decay speedily. So an unnecessarily large number of data points only decreases the S/N ratio of the signals. Another problem caused by a long acquisition time is that it causes the use of a long repetition time and increases the time of measurement. So it is important to choose the proper condition of measurement for achieving adequate digital resolution. The digital resolution can also be improved by reducing the spectral width. In this case the folded peaks appear as usual (see Chap. 1.2), when the peaks lie outside the spectral range. This is a point to note.

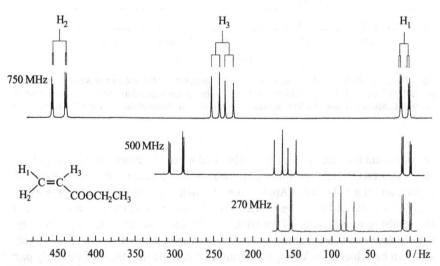

Fig. 2.8. ^{1}H NMR spectra of ethyl acrylate at different frequences. 5 wt/vol% CDCl$_3$ solution, 30 °C, 45° pulse, pulse repetition time 20 s, 16 scans

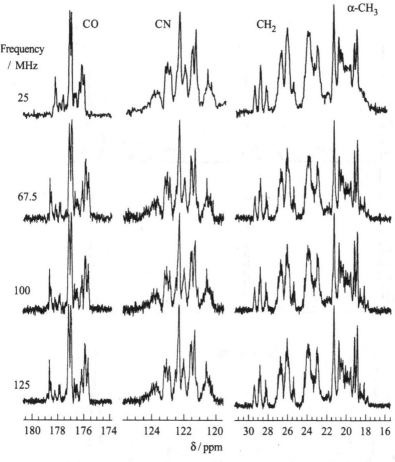

Fig. 2.9. ^{13}C NMR spectra of copolymer of MMA and acrylonitrile (*AN*) (MMA:AN=34.7:65.3 mol/mol) measured at different frequencies [2]. 10 wt/vol% CD$_3$CN solution, 70 °C, 45° pulse, pulse repetition time 3 s, 12,000–24,000 scans

Separation of signals increases as the applied magnetic field increases. The ^{1}H NMR signals of methylene and methine protons of ethyl acrylate measured at 270, 500, and 750 MHz are shown in Fig. 2.8. The dependence of the chemical shift on the applied magnetic field and the independence of the spin–spin coupling constant enable us to distinguish between them. The carbonyl carbon signals of poly-(MMA-*co*-acrylonitrile) with 34.7:65.3 monomeric units are shown in Fig. 2.9. The increased peak separation of the signals due to the comonomer sequence distribution is much more clearly observed in the spectrum taken at 125 MHz [2, 14].

Peak separation also depends on the temperature of measurement and the solvent used. The ^{1}H NMR spectra of PMMA recorded in benzene-d_6 at different temperatures are shown in Fig. 2.10. Peak separation and resolution are greatly

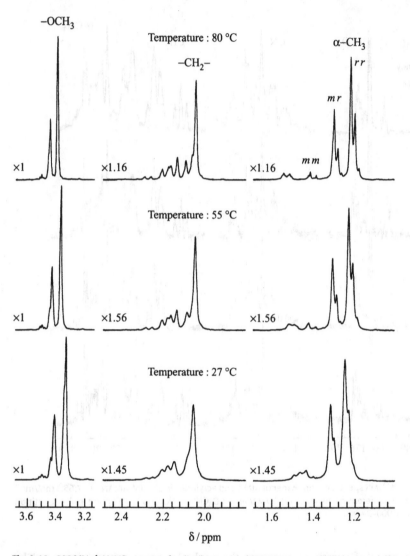

Fig. 2.10. 500 MHz ^{1}H NMR spectra of radically prepared PMMA measured in benzene-d_6 (10 wt/vol%) at different temperatures [2]. 45° pulse, pulse repetition time 10 s, 16 scans

enhanced at higher temperatures. Particularly, the peak separations in the signals of α-CH$_3$ and CH$_2$ groups, whose segmental mobilities are low compared with the CH$_3$O– group, are greatly improved and the separations due to the pentad and tetrad tacticities become clear as the temperature increases. As mentioned in Sect. 1.1.1 the intensities of the signals decrease with increasing temperature of measurement; however, the narrowing of the peak at high temperatures usually cancels the effect of the decrease in signal intensity, and in the case of a polymer the S/N ratios are enhanced at high temperatures except for very high temperatures. The extent of the peak separation changes also with the solvents used as mentioned in Sect. 1.4, Fig. 1.15.

^{1}H NMR spectra should always be quantitative if measured under appropriate conditions. The values of the repetition time and the flip angle should be carefully chosen as mentioned in Figs. 2.1 and 2.2. Large flip angles increase the signal intensities but require long repetition times to avoid signal saturation. Small flip angles enable the application of short repetition times but decrease the signal intensities. Therefore optimization is needed to obtained the ^{1}H NMR spectrum with high S/N ratio and quantitativeness. Quantitative determination with ^{1}H NMR spectra is also described in Sects. 2.2–2.4.

2.2 Reliability of Chemical Shift and Signal Intensity

As described in the previous section, accuracy and precision is of particular importance in NMR measurements of polymers. In order to assess the reliability of NMR data, the Research Group on NMR, the Society of Polymer Science, Japan (SPSJ), collected ^{1}H NMR and ^{13}C NMR spectra of two identical samples, a radically prepared PMMA and solanesol (nonamer of isoprene) from a number of NMR spectrometers by a round-robin method [13].

^{1}H NMR spectra of the PMMA in CDCl$_3$ were measured on 26 spectrometers, whose resonance frequencies ranged from 90 to 500 MHz. The chemical shift data of methoxy protons scattered only about ±0.01 ppm, with a few exceptions. The standard deviation was 0.0038 ppm for 19 data obtained from FT instruments and 0.0169 ppm for seven data obtained from CW ones. Standard deviations for α-methyl and methylene proton shifts were less than 0.01 ppm.

The relative intensity data, 3(CH$_2$+α-CH$_3$)/OCH$_3$, agreed well with the theoretical value of 5. Standard deviations for the intensity measurements were larger for the FT method (3.2%) than for the CW method (1.1%), while those for chemical shifts were larger for the CW method. The precision of the intensity measurement by a single FT NMR spectrometer operated at 500 MHz was higher (less than 2%) than that observed by the round-robin method.

The ^{13}C NMR spectra of the same PMMA sample in CDCl$_3$ were measured on 24 instruments, whose resonance frequencies ranged from 15.0 to 125 MHz. All carbons except for the methoxy one are sensitive to the tacticity of the polymer chain. The standard deviations for the ^{13}C chemical shift measurements were within ±0.05 ppm. The standard deviations for methoxy and CDCl$_3$ carbons were small compared with those for the others. The measurements for solanesol, whose molecular weight is lower than that of PMMA, were a little more precise than those for PMMA.

The relative peak intensities for carbonyl, (methoxy+methylene), and quaternary carbons against the α-methyl carbon gave much poorer accuracy than those of ^{1}H NMR (Table 2.5) [13, 15]. Only six instruments (22.5 and 25 MHz instruments) out of 14 were found to give data close to the theoretical values in all three relative intensities within ±15% deviation under complete ^{1}H decou-

Table 2.5. Relative intensities of ^{13}C NMR signals of PMMA in $CDCl_3$ at 55 °C (A=C=O/ α-CH_3, B=(CH_2+OCH_3)/α-CH_3, C=C-4/α-CH_3. Pulse width 45–90°, pulse repetition time 1–20 s) [13]

Frequency (MHz)	Complete decoupling				Complete decoupling without NOE[a]			
	A	B	C	n[b]	A	B	C	n[b]
22.5	0.86	1.93	1.00	3	1.12	2.21	1.19	2
25	0.86	2.00	1.00	3	0.97	2.00	1.03	1
50	0.54	1.45	0.92	3	0.89	2.00	1.03	1
67.5	0.55	1.52	0.75	1	0.99	1.95	1.02	2
90.6	0.43	1.33	0.70	1	–	–	–	–
100	0.33	1.54	0.67	1	1.00	2.06	1.08	2
125	0.48	1.58	0.83	1	0.94	2.10	1.18	3

[a] Measurement with gated decoupling without NOE (pulse repetition time 20–25 s).
[b] Number of determinations.

pling conditions. The values of the relative intensities obtained with instruments at 50 MHz or higher frequencies generally showed large negative deviations. One of the sources of errors in these quantitative measurements is the variation in the NOE values (Sects. 7.1, 7.2). The NOE values for each carbon at 22.5 and 25 MHz were close to 2.99. However, the NOE decreases with increasing magnetic field strength strongly depending on the type of carbon (Sect. 7.5) [15]. As described in the previous section (Table 2.1), the measurement on a 125 MHz spectrometer at a pulse repetition of 5 s with a 45° pulse gave reliable relative intensities with 2–9% accuracy and about 2% precision when the spectra were measured under a gated decoupling without NOE. The relative intensities obtained with gated decoupling through the round-robin method are also close to the theoretical values as shown in Table 2.5.

The high performance of modern NMR spectrometers enables us to measure the ^{13}C NMR spectra of high S/N ratios without ^1H-^{13}C spin decoupling. The signal intensities of the spectrum for radically prepared PMMA measured without the decoupling agreed well with the theoretical values. The problem in this case is the complicated overlap of two resonances that are split into several peaks by ^{13}C-^1H spin–spin coupling (Table 2.1).

2.3   Determination of Absolute Signal Intensity

Quantitative analysis by NMR usually deals with the determination of the relative intensities of the signals. In the quantitative analysis, in which the absolute intensity of the signal of interest is required, external [16, 17] or internal [18, 19] standard methods can be used. Determination of the hydrogen content of the sample by

Table 2.6. Internal standards for quantitative determination [19]

Compound	Formula	δ in $CDCl_3$ (ppm)
Hexamethyl-cyclotrisiloxane		0.168
Trioxane		5.10
1,1,2,2-Tetra-chloroethane	$HCCl_2-CHCl_2$	5.95
Trichloroethylene	$Cl_2C{=}CHCl$	6.461
Pyrazine		8.51

the external standard method can be carried out by using two NMR sample tubes with highly precise dimensions. One of the tubes is filled with a standard solution of a known compound with a known concentration, and the other with a solution of the sample of interest with a known concentration. The NMR spectra of the two solutions are measured alternately under the same conditions and the hydrogen content of the sample is determined by comparing the signal intensities of the sample with those of the standard [16, 17]. The problem in this method is the errors due to the instrumental variation during the measurements of the spectra, and it is rather difficult to obtain good results.

Absolute quantitative analysis can also be made by using an internal standard, which is added to the solution of the sample of interest in a certain amount. The internal standard must be chosen so that it does not conflict with the sample of interest; it should be chemically inert and nonvolatile and should give a sharp singlet signal in a clear region of the spectrum of the sample solution. Some of the internal standards selected in this sense are shown in Table 2.6 [19] together with their chemical shifts.

Use of a pair of internal standards is recommended to avoid errors due to the decomposition of one standard during the measurement, inaccuracies due to overlap of peaks from one internal standard with small impurities, and errors due to carelessness [19].

The internal standard method has an advantage over the external standard method in that random variations during the measurements are cancelled out completely. However, the contamination of the sample by the standard compound and the troublesome work to mix a reasonable amount of the standard with the sample solution are disadvantages of the internal standard method.

Problems due to the instrumental variation in the external standard method and contamination of the sample by the standard compound in the internal standard method can be avoided by the technique for quantitative analysis with precision coaxial tubing [20, 21]. The coaxial tubing illustrated in Fig. 1.18b is used for this technique. The intensity standard is placed in the central capillary and the solution of the sample of interest in the surrounding annulus. One of the typical intensity standards is a solution of trimethylaluminum in toluene-d_8; it shows a singlet signal at about 1 ppm higher field than the signal of TMS and is therefore preferable for the analysis of various types of organic compounds. The singlet is not too sharp and not too broad, and is favorable for accurate measurements of intensity.

The ratio, R, of the signal intensity of the sample to that of trimethylaluminum is determined. The hydrogen content of the sample is calculated according to the following equation:

$$H(\%) = 100\, F \times R/c. \tag{2.3}$$

Here, c (grams per liter) is the concentration of the sample and F (grams of hydrogen per liter) is the concentration factor of the standard solution in the inner cell. To determine the value of F, the inner cell containing the intensity standard was assembled with the outer cell containing a known amount of toluene in carbon tetrachloride. In this coaxial tubing method, the signal intensity of the intensity standard can be adjusted to be close to that of the sample of interest by choosing the inner cell containing an appropriate amount of intensity standard in order to enhance the accuracy of determination. The accuracy is usually higher and the time required for the analysis is shorter for the coaxial tubing method than for the external or internal standard method mentioned previously.

Using an inner cell of lower F value, the amounts of a nondeuterated hydrogen compound in deuterated solvents were accurately and precisely determined as shown in Table 2.7 [2]. The contents of the nondeuterated solvents were within the guaranteed range for all the deuterated solvents. The contents of water in the solvents were also determined. Some of the solvents were analyzed by Karl–Fisher titration and the results agreed well with those obtained by the NMR method. The results naturally lead to the possibility of microanalysis by the coaxial tubing method as described in Sect. 2.4.

The coaxial tubing method is also powerful in the analysis of multicomponent mixtures. The waste solvents in our laboratory were analyzed by 100 MHz ^{1}H NMR using several inner cells containing different amounts of the standard compound [21]. The results are shown in Table 2.8. The calculations of the contents of carbon disulfide and carbon tetrachloride were carried out from the results of elementary analysis by assuming that the former was the only sulfur-containing compound in the solvent and that the latter was the only chlorine-containing compound without

Table 2.7. Degree of deuteration and water content of commercial deuterated solvents determined by coaxial tubing method [2]

Solvent	Company	Degree of deuteration (%)		Water (ppm)
		Guaranteed	Observed	
Chloroform-d	CEA	99.96	99.97	26.3
	CEA	99.96	99.96	26.3
	Merck	99.8	99.85	104.6
	Merck	99.7	99.81	113.1
	Aldrich	99.8	99.86	232.1
Acetone-d_6	Aldrich	99	99.45	22.5
Methanol-d_4	Aldrich	99.5	99.53	–
Ethanol-d_6	Merck	99	99.36	307.4
Dichloromethane-d_2	CEA	99.3	99.59	40.7
Nitromethane-d_3	Merck	99	99.08	449.3
Acetonitrile-d_3	Merck	99	99.27	283.6
DMSO-d_6	CEA	99.8	99.81	218.2
Benzene-d_6	Aldrich	99.5	99.63	64.4
Toluene-d_8	Aldrich	99	99.41	97.5
Nitrobenzene-d_5	Merck	99	99.74	117.6

Table 2.8. Composition of a waste solvent. Elementary analysis: C 76.04%, H 9.89% (combustion method); C 78.23%, H 10.25% (calculated from the NMR results) [21]

Compound	wt/wt%	Compound	wt/wt%
Carbon disulphide	0.35	Dichloromethane	1.48
Benzene	18.57	Cyclohexane	0.26
Toluene	20.45	Methanol	1.14
n-Hexane	23.80	Acetone	3.34
Diethyl ether	4.01	Carbon tetrachloride	3.19
Ethyl acetate	0.62	Ethanol	1.79
Methyl acetate	0.54	Trichloroethylene	1.03
Water	0.40	Vegetable oil	16.44
THF	0.18	Unknowns	2.24
Dioxane	0.17		

hydrogen. The content of vegetable oil was evaluated from the characteristic NMR signal at 6.2 ppm (triplet). The fraction of unknowns is the difference between the sum total of defined components and 100. The carbon and hydrogen contents in the solvent calculated from the NMR data are in agreement with those obtained by elemental analysis (Table 2.8).

2.4 Microanalysis by NMR Spectroscopy

Analysis of end groups or abnormal linkages in polymer chains often gives us important information on the mechanism of polymerization (see Sects. 5.1, 5.2). Although the low concentrations of these groups have often made it difficult to analyze, a superconducting NMR spectrometer permits the measurements of spectra with greater sensitivity and thus more detailed analysis of microstructures in polymers. Beside the microstructure analysis, quantitative determination of samples or species with low concentrations is often necessary for studies in the fields of polymer chemistry.

As described in the previous section, the coaxial tubing method is applicable to the quantitative analysis of the sample of interest with a wide range of concentration by selecting an inner cell with a proper F value, and is naturally useful in microanalysis. Combined with a 500 MHz ^{1}H NMR spectrometer, the method allowed us to make a quantitative analysis for very small amounts of impurities contained in the solvents used for NMR measurements. Commercial carbon tetrachloride was found to contain small amounts of benzene, chloroform, and dichloromethane in 1.03, 37.85, and 1.49 ppm (wt/vol), respectively. The inner cell used for the analysis contained HMDS in acetone-d_6 as the intensity standard and the factor F was 1.43×10^{-5} g hydrogen/l. With 0.5-h measurement, benzene or dichloromethane in the order of 1 ppm could be analyzed quantitatively. To examine the lower limit of the analysis, the sample was diluted 20 times with acetone-d_6 and analyzed by the same procedure. The results shown in Table 2.9 indicate that the quantitative determination for the order of 10^{-2} ppm of sample is possible with good accuracy [22].

In the microanalysis by ^{1}H NMR spectroscopy, impurities contained in NMR solvents, even at very low levels, may give interfering signals. The 500 MHz ^{1}H NMR spectrum of chloroform-d is illustrated in Fig. 2.11 [22]. Along with strong signals due to undeuterated chloroform and water, signals due to impurities in the amount of a few ppm or less are observed. The sample was taken from the container of chloroform used for daily work in our laboratory. Among the impurities, haloalkanes may be contained originally in the solvent, while TMS, HMDS, and

Table 2.9. Quantitative determination of C_6H_6, $CHCl_3$, and CH_2Cl_2 in $CCl_4/(CD_3)_2CO$- (1/19 vol/vol) by a precision coaxial tubing method [45° pulse, pulse repetition time 10 s, 8,411 scans, 23.5 h, inner cell HMDS/$(CD_3)_2CO$] [22]

Compound	Concentration (ppm)		Accuracy (%)
	Taken	Found	
C_6H_6	0.052	0.047	9.6
$CHCl_3$	1.890	1.792	5.3
CH_2Cl_2	0.074	0.077	4.1

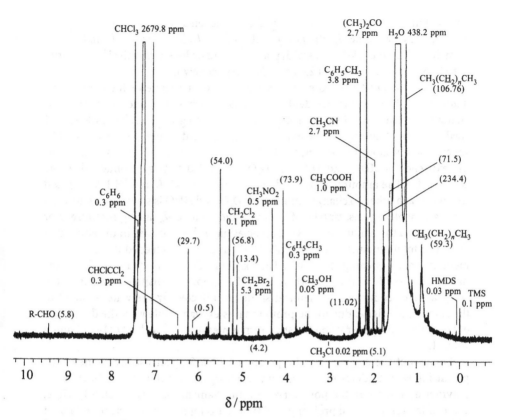

Fig. 2.11. 500 MHz ^{1}H NMR analysis of chloroform-d at 30 °C. The amounts of known compounds are expressed in ppm (wt/vol), and those of unknown impurities are shown in *parentheses* in units of 10^{-6} g hydrogen/l [22] 45° pulse, pulse repetition time 20 s, 1,600 scans

acetone may be introduced into the solvent during the laboratory manipulation. It is advisable in the microanalysis of a series of polymer samples to use solvents well-characterized for impurities at the ppm level.

Water is often a major impurity in deuterated solvents, especially in polar solvents. For example, a commercial acetone-d_6 contained about 250 ppm of water ($H_2O+HDO+D_2O$) as determined by ^{1}H NMR and ^{2}H NMR spectroscopies even immediately after opening the bottle. The water content in the acetone-d_6 kept in an NMR sample tube with a plastic cap increases at the rate of 110 ppm/day when allowed to stand in ordinary chemical laboratories. When the sample in the NMR tube is kept in a finely air-conditioned room, such as the rooms used to house NMR spectrometers, the rate of water uptake decreases to 40 ppm/day. Sealing with PARAFILM did not affect the rate. Sealing with a heat-shrinkable polyethylene tube (Fig. 1.22) effectively decreases the rate to around 10 ppm/day. When stored in a refrigerator, the rate of the water uptake for the tube with a plastic cap is 5.4 ppm/day. The water content does not increase when stored in a reagent bottle tightly capped and kept in a refrigerator. The water uptake rate depends on the solvent (Sect. 1.4.1) and the rates for actone-d_6, ethanol-d_6, chloroform-d, nitrobenzene-d_5, and carbon tetrachloride in the NMR tubes with plastic caps in ordinary chemical laboratories are 110.6, 40.1, 39.3, 37.7, and 2.4 ppm/day, respectively [22].

In the course of these experiments, we noticed that some of the commercial deuterated solvents contained deuterium oxide, which in turn underwent H–D exchange with H_2O introduced during storage or use to give HDO. ^1H NMR and ^2H NMR spectra of commercial acetone-d_6 are illustrated in Fig. 2.12. The ^1H NMR spectra clearly show the existence of H_2O and HDO and the ^2H NMR spectra show the existence of HDO and D_2O, i.e., D_2O, HDO, and H_2O are contained in the acetone-d_6.[2] The amounts of HDO determined by ^1H NMR and ^2H NMR agreed well with each other. Exchange among H_2O, D_2O, and HDO is slow enough to show distinguishable signals. Uptake of H_2O into the acetone-d_6 during the storage or use increased the amount of HDO in association with a decrease in the amount of D_2O. The total amount of water (D_2O, HDO, and H_2O) exceeded 0.3% even immediately after opening the bottle. The existence of D_2O in deuterated solvents may cause trouble in the case where an NMR sample is sensitive to water or the spectral features including chemical shifts are affected by the water content in the sample; the signal due to the proton that is exchangeable with the deuterium of D_2O might disappear when the spectrum is measured in such a D_2O-containing solvent.

Analysis of the end groups of a polymer by NMR spectroscopy is one of the promising ways to determine the number-average molecular weight ($\bar{M}_n$) of the polymer and to study the polymerization mechanism (see also Sect. 5.1). Here, some of the fundamental problems and typical examples for the determination of $\bar{M}_n$ by NMR spectroscopy are described. When the chemical structure of a polymer including end groups is definitely known as $A-(M)_n-B$ (M is a monomeric unit), the quantitative determination of the relative intensities of the end groups to the monomeric units is a simple and straightforward method to provide the value of $\bar{M}_n$. Such a well-defined polymer sample can be obtained by living polymerization, in which the polymerization proceeds without chain transfer and termination reactions and every initiating species produces one polymer chain that has one initiator fragment at the chain end. Polymerization of MMA with t-C_4H_9MgBr in toluene at low temperature proceeds in a living manner to afford highly isotactic PMMA with the following structure [23, 24].

$$\begin{array}{cc} CH_3 & CH_3 \\ | & | \\ CH_3-C-(CH_2-C)_n-H \\ | & | \\ CH_3 & C=O \\ & | \\ & OCH_3 \end{array} \qquad (2.4)$$

---

[2] The acetone-d_6 made by Merck in the early 1970s did not contain D_2O and HDO, nor did the sample stored for at least for 18 years, indicating that spontaneous D–H exchange between acetone-d_6 and H_2O does not occur.

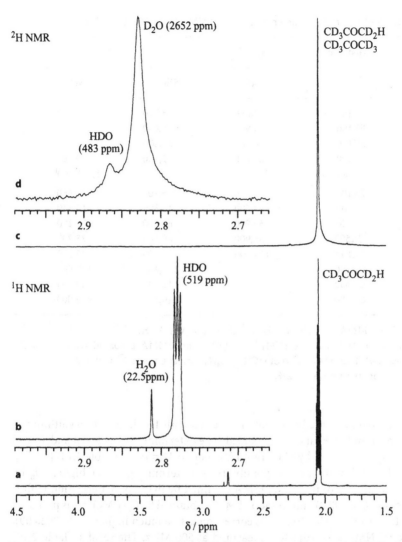

Fig. 2.12. a,b 270 MHz ^{1}H NMR and **c,d** 41.34 MHz ^{2}H NMR spectra of acetone-d_6 (99% D) at 30 °C [22] 45° pulse, pulse repetition time: 20 s, 128 scans **a** and 1080 scans **c**

The ^{1}H NMR spectrum of PMMA in nitrobenzene-d_5 measured at 110 °C showed a singlet signal of the t-C_4H_9 protons at 0.82 ppm separately from α-CH_3 proton signals. The intensity of the t-C_4H_9–signal, $I(t$-$C_4H_9)$, against that of the monomeric units, for example, the OCH_3 proton signal at 3.60 ppm, $I(OCH_3)$, provides $\bar{M}_n$:

$$\bar{M}_n = [3 \times I(OCH_3)/I(t\text{-}C_4H_9)] \times 100.14 + 58. \tag{2.5}$$

The $\bar{M}_n$ values thus determined by 100 MHz ^{1}H NMR spectroscopy agreed well with those determined by size-exclusion chromatography (SEC) and vapor pressure osmometry (VPO) up to the $\bar{M}_n = 2 \times 10^4$ (Table 2.10) [22, 24]. $\bar{M}_n$ of syndiotactic

Table 2.10. Molecular-weight determination of PMMA by ^{1}H NMR [22, 24]

PMMA[a]	Instrument	$\bar{M}_n$		
		VPO	SEC	NMR
A	FX100	3,660	3,510	3,560
	FX100	4,930	5,650	4,940
	FX100	10,100	10,400	9,520
	FX100	21,200	21,200	20,800
	GX500			1,891,000[b]
B	FX100	5,200	5,100	5,660
	FX100	12,600	10,300	13,340
	FX100	55,900	60,300	75,100
	GX500	55,900	60,300	75,200
	FX100	119,400	119,400	132,300
	GX500	119,400	119,400	138,000[c]
	GX500		217,900	249,500
	GX500		610,300	455,000

[a] A=t-C$_4$H$_9$-(-MMA-)$_n$-H. B=n-C$_5$H$_{11}$C(C$_6$H$_5$)$_2$-(-MMA-)$_n$-H.
[b] A mixture of PMMA-A [$\bar{M}_n$ (NMR) = 20,800] and PMMA prepared with C$_6$H$_5$MgBr (1/93.35 w/w). The proton ratio of OCH$_3$/t-C$_4$H$_9$ corresponds to $\bar{M}_n$ = 1,962,000.
[c] Precision for six runs was 3.02%.

PMMAs prepared by the living polymerization with 1,1-diphenylhexyllithium in tetrahydrofuran (THF) at –78 °C could also be determined by ^{1}H NMR and agreed with the $\bar{M}_n$ determined by SEC and/or VPO up to $\bar{M}_n$=5×10^5 (Table 2.10) [22, 24].

In order to investigate the possibility of determining much higher $\bar{M}_n$ by ^{1}H NMR, a mixture of isotactic PMMA obtained by living polymerization with t-C$_4$H$_9$MgBr ($\bar{M}_n$=20,800) and isotactic PMMA obtained by C$_6$H$_5$MgBr was prepared so that the ratio of OCH$_3$ to t-C$_4$H$_9$ corresponds to a much higher $\bar{M}_n$ (1,962,000), and the ^{1}H NMR spectrum was measured at 500 MHz. The results (Table 2.10) indicate that the intensity ratio corresponding to $\bar{M}_n$=2×10^6 can be accurately determined by 500 MHz ^{1}H NMR spectroscopy [22].

The ^{1}H T_1 for the signals of the end groups are usually longer than those for the signals due to the in-chain group of polymers. This should be carefully considered in the end-group analysis; the pulse repetition time should be chosen considering primarily the values of T_1 for the end-group signals. Another important point that should be considered is that the T_1 for the signal of the in-chain group increases with an increase in the resonance frequency, while the T_1 for the signal of the end group is less sensitive to the frequency. For example, the T_1 (1.06 s) for the signal of the t-C$_4$H$_9$- end group of isotactic PMMA is 1.7 times as long as the T_1 (0.64 s) for the in-chain OCH$_3$ when measured at 100 MHz in nitrobenzene-d_5 at 110 °C, while both the T_1 (1.15 and 1.21 s, respectively) are almost equal when measured

Table 2.11. ^{1}H NMR determination of a small amount of CH_3NO_2 in nitrobenzene-d_5 solution of PMMA (values indicate intensity ratio of OCH_3/CH_3NO_2) [22]. The values in *parentheses* express accuracy in percentage

Taken	Found[a]	
	100 MHz (ADC 12 bit)	500 MHz (ADC 16 bit)
46.8	47.8 (2.1)	49.3 (5.3)
210.0	199.8 (4.9)	204.7 (2.5)
913	897 (1.8)	950 (4.0)
6,940	7,220 (4.0)	7,250 (4.5)
14,791	13,040 (11.8)	14,200 (3.9)
30,401		27,800 (8.7)
61,649		67,900 (10.2)
250,337		228,000 (8.9)

at 500 MHz in nitrobenzene-d_5 at 110 °C. This indicates that measurement at a higher magnetic field is generally preferable for the $\bar{M}_n$ determination by NMR.

In such an NMR measurement where weak peaks should be observed in the presence of extremely strong peaks due to monomeric units, the dynamic range of the receiver amplifier of the spectrometer as well as that of the ADC come into question. To test this problem experimentally for polymer samples, nitrobenzene-d_5 solutions of PMMA containing small amounts of nitromethane (CH_3NO_2) were prepared and the ratio of the peak intensity for CH_3NO_2 and OCH_3 of PMMA was measured on 100 and 500 MHz ^{1}H NMR spectrometers, to which 12-bit and 16-bit ADCs were installed, respectively (Table 2.11). When the 100 MHz spectrometer (12 bit) was used, the amount of CH_3NO_2 could be determined with good accuracy up to an intensity ratio of CH_3NO_2 to OCH_3 (PMMA) of 1/6,940 and was measurable up to a ratio of 1/14,791. With the 500 MHz spectrometer (16 bit) the measurable ratio was as small as 1/250,000.[3] The time required for the latter measurement was 1 day. The result suggests that the $\bar{M}_n$ of PMMA can be determined up to about 25 million ($\overline{DP}$=250,000) as far as a polymer molecule contains exactly one terminal CH_3 group that shows a signal apart from the strong signals due to monomeric units.

^{13}C NMR spectroscopy may afford a much better chance to observe end-group signals separately from main-chain signals; however, the 1.1% natural abundance of ^{13}C in carbon compounds and the low concentration of the end groups make the

---

[3] The methoxy proton signal of PMMA is multicomponent owing to the tactic pentads and is much broader than the signal of CH_3NO_2, and thus the measurable intensity ratio of the whole OCH_3 signal against that of CH_3NO_2 is larger than the dynamic range expected from the bit number of the ADC [12 bits 2^{12}=4,096 (100 MHz); 16 bits 2^{16}=65,536 (500 MHz)].

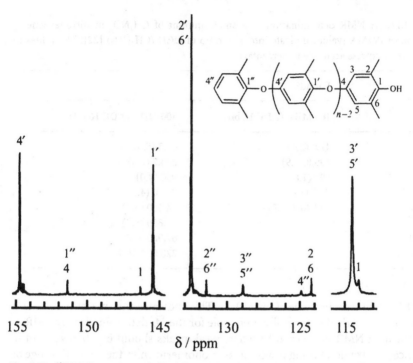

Fig. 2.13. ^{13}C NMR spectrum of oligo(oxy-2,6-dimethyl-1,4-phenylene) measured at 24 °C and at 90.56 MHz [25]. 70° pulse, pulse repetition time 5 s

quantitative analysis difficult. One approach is the use of a ^{13}C-labeled initiator and the preparation of end-labeled polymers as described in Sect. 5.1. In the case of a relatively low molecular weight polymer, $\bar{M}_n$ can be determined rather easily by ^{13}C NMR spectroscopy. A typical example is the determination of $\bar{M}_n$ (1,600–3,200) of poly(oxy-2,6-dimethyl-1,4-phenylene) [25]. The ^{13}C NMR spectra measured in CDCl$_3$ is shown in Fig. 2.13. The peak assignment indicated in the figure was made with the aid of a lanthanide shift reagent Pr(fod)$_3$ (Sect. 1.4.4). For the $\bar{M}_n$ determination the spectral region 145–155 ppm is chosen, in which the signals due to the nonprotonated carbons bonded to oxygen appear, and accordingly the complications due to unequal NOEs are effectively eliminated. The values of $\bar{M}_n$ thus obtained ranged from 1,600 to 3,200 and the reciprocal of the $\bar{M}_n$ showed a linear correlation with the glass-transition temperature of the polymer, indicating the reasonableness of the $\bar{M}_n$ determination [25].

The limit of detection of end groups by ^{13}C NMR also depends on the structure. The t-butyl group is a favorable end group for observation since it consists of three equivalent methyl carbons which show a sharp singlet. The ^{13}C NMR signal of the t-C$_4$H$_9$ group in the isotactic PMMA with $\bar{M}_n$=11,600 as determined by ^{1}H NMR (Eq. 2.5) could be observed with a S/N ratio of 3.8 when the spectrum was measured in nitrobenzene-d_5 at 100 °C under the gated decoupling condition (concentration 10 wt/vol%, 3,600 scans, pulse repetition time 20 s) [26]. The relative intensities of the signals due to CH_2 and α-CH_3 to CH_3 of t-C_4H_9 group corresponded to $\bar{M}_n$ values of 11,960 and 12,030, respectively, which are consistent with the $\bar{M}_n$ determined by ^{1}H NMR.

Combination of rapid and quantitative reaction of the end groups with a certain reagent and NMR analysis of the resulting polymer is sometimes very useful for the $\bar{M}_n$ determination. The reactions of hexafluoroacetone (HFA) with primary and secondary hydroxy end groups of polyethers such as poly(oxyethylene) and poly(oxypropylene) are very rapid and quantitative. ^{19}F NMR analysis of the resulting HFA adducts of the polymers provides accurate information of the $\bar{M}_n$ and the structural environment of the hydroxy end group. n-Butyl trifluoroacetate is usually used as an intensity standard for the ^{19}F NMR analysis. The measurement is not interfered with the existence of water, acids, amines, and amides, which is an important advantage over the method based on chemical titration [27].

$$(CF_3)_2C = O + ROH \rightarrow (CF_3)_2C(OR)(OH) \tag{2.6}$$

2.5 Determination of Volume Magnetic Susceptibility by NMR

Several methods were reported to determine the volume magnetic susceptibility (χ_v) by ^{1}H NMR using the coaxial tubing with a CW spectrometer in which the direction of magnetic field is perpendicular to the axis of sample spinning [28–31]. These are easy and elegant ways to determine the χ_v of organic compounds; but, very unfortunately, CW spectrometers are not available any more in most of chemical laboratories.

The chemical shifts in ppm of the sample in the outer cell of coaxial tubing (Fig. 1.18b) observed with a superconducting solenoid spectrometer, δ_1, and an electromagnet spectrometer, δ_2, by referring the shifts to the external standard signal from the inner cell are different from each other [32]. The difference between the chemical shifts, δ_1–δ_2, can be described as follows:

$$\delta_1 - \delta_2 = 2\pi\,(\chi_s - \chi_r), \tag{2.7}$$

where χ_s and χ_r represent the χ_v of the sample in the outer cell and that of the reference sample in the inner cell. By using the known χ_r value of the reference, the χ_s value of the sample can be determined according to Eq. (2.7) [32, Terawaki Y, Kitayama T, Hatada K (unpublished results)].

By using the modified Frei–Niklaus microcell (Fig. 1.18c) and a superconducting solenoid spectrometer, χ_v values can be determined easily as follows [Terawaki Y, Kitayama T, Hatada K (unpublished results), 33, 34]. The sample solution is placed in the inner cell and the reference compound, for example, $CDCl_3$, in the space around the spherical cell. The deuterium signal from the $CDCl_3$ is used for NMR locking. The depth of the sample tube is adjusted so that the sphere (bulb) of the inner cell is placed at the center of the detection coil. Then the observed ^{1}H NMR spectrum of the sample solution shows two peaks for each resonance, one

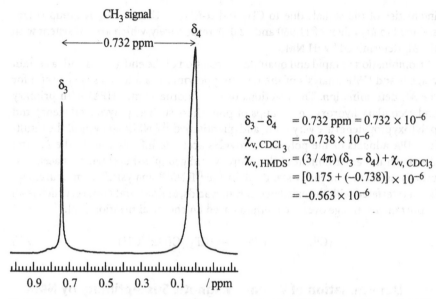

Fig. 2.14. Determination of the χ_v for HMDS' using a modified Frei–Niklaus microcell and a 500 MHz NMR spectrometer at 25 °C [33]. Inner cell HMDS', outer cell CDCl$_3$

from the sample in the bulb, δ_4, and the other from the attached stem, δ_3. The chemical shifts between these two peaks are proportional to the difference in the χ_v between the sample solutions in the inner cell and the reference solution in the outer one. Then the χ_v value of the sample solution can be determined according to Eq. (2.8). A typical example of the determination for hexamethyldisilane (HMDS') is shown in Fig. 2.14.

$$\delta_3 - \delta_4 = \frac{4}{3}\pi\left(\chi_s - \chi_r\right) \tag{2.8}$$

References

1. HATADA K, KITAYAMA T, UTE K (1993) Annu Rep NMR Spectrosc 26:99
2. HATADA K, KITAYAMA T, NISHIURA T, TERAWAKI Y (1995) Kobunshi-Jikkengaku 5:23
3. DEROME AE (1987) Modern NMR techniques for chemistry research. Pergamon, New York
4. HATADA K, TERAWAKI Y (1968) Kogyo Kagaku Zasshi 71:1163
5. HATADA K, OTA K, TERAWAKI Y, YUKI H (1968) Kogyo Kagaku Zasshi 71:1168
6. HATADA K, TERAWAKI Y, OKUDA H, NIINOMI S, YUKI H (1971) Kobunshi Kagaku, 28:293
7. HATADA K, TERAWAKI Y, OHSHIMA J, YUKI H (1972) Kobunshi Kagaku 29:391
8. MÜLLEN K, PREGOSIN PS (1976) Fourier transform NMR techniques: a practical approach. Academic, New York
9. HOULT DH (1979) In: Levy GC (ed) Topics in carbon-13 NMR spectroscopy, vol 3. Wiley, New York, p 16

10. (a) THIAULT B, MERSSEMAN M (1975) Org Magn Res 7:575; (b) THIAULT B, MERSSEMAN M (1976) Org Magn Res 8:28

11. RANDALL JC (1977) Polymer sequence determination – Carbon-13 NMR method. Academic, New York, chap 5

12. HATADA K, TERAWAKI Y, KITAYAMA T, UTE K (1989) Polym Prepr Jpn 38:845

13. CHÛJÔ R, HATADA K, KITAMARU R, KITAYAMA T, SATO H, TANAKA Y, members of research group on NMR SPSJ (1987) Polym J 19:413

14. HATADA K, KITAYAMA T, TERAWAKI Y, SATO H, CHÛJÔ R, TANAKA Y, KITAMARU R, ANDO I, HIKICHI K, HORII F, members of research group on NMR SPSJ (1995) Polym J 27:1104

15. CHÛJÔ R, HATADA K, KITAMARU R, KITAYAMA T, SATO H, TANAKA Y, HORII F, TERAWAKI Y, members of research group on NMR SPSJ (1988) Polym J 20:627

16. SMITH WB (1964) J Chem Educ 41:97

17. WILLIAMS RB (1958) Ann NY Acad Sci 70:890

18. BARCZA S (1963) J Org Chem 28:1914

19. COCKERILL AF, HARDEN RC, DAVIES GLO, RACKHAM DM (1974) Org Magn Reson 6:452

20. HATADA K, TERAWAKI Y, OKUDA H, NAGATA K, YUKI H (1969) Anal Chem 41:1518

21. HATADA K, TERAWAKI Y, OKUDA H (1977) Org Magn Res 9:518

22. HATADA K, TERAWAKI Y, KITAYAMA T (1992) Kobunshi Ronbunshu 49:335

23. HATADA K, UTE K, TANAKA K, OKAMOTO Y, KITAYAMA T (1985) Polym J 17:977

24. HATADA K, UTE K, TANAKA K, OKAMOTO Y, KITAYAMA T (1986) Polym J 18:1037

25. REUBEN J, BISWAS A (1991) Macromolecules 24:648

26. HATADA K, KITAYAMA T, UTE K, TERAWAKI Y, FUJIMOTO N (1991) Polym Prepr Jpn 40:1119

27. Ho FF-L (1973) Anal Chem 45:603

28. ZIMMERMAN JR, FOSTER MR (1957) J Phys Chem 61:282

29. BERNSTEIN HJ, FREI K (1962) J Chem Phys 37:1891

30. HATADA K, TERAWAKI Y, OKUDA H (1969) Bull Chem Soc Jpn 42:1781

31. HATADA K, TERAWAKI Y, OKUDA H (1972) Bull Chem Soc Jpn 45:3720

32. BECCONSALL JK, DOYLE DAVES G JR, ANDERSON WR (1970) J Am Chem Soc 92:430

33. MOMOKI K, FUKAZAWA Y (1990) Anal Chem 62:1665

34. MOMOKI K, FUKAZAWA Y (1994) Anal Sci 10:53

3 Stereochemistry of Polymers

When a polymer has stereochemical isomerism within the chain, its properties often depend on the stereochemical structure. Thus the analysis of the stereochemistry of polymers is important and NMR spectroscopy has been the most valuable tool for this purpose.

3.1 Definition of Tacticity

In polymers of vinyl monomers $CH_2=CH-X$ or vinylidene monomers $CH_2=CXY$, the main-chain carbons having substituent group(s) are termed "pseudo-asymmetric" since, if the chain ends are disregarded, such carbons do not have the four different substituents necessary to qualify for being truly asymmetric. Nevertheless, they have the possibility of relative handedness. The simplest regular arrangements along a chain are the isotactic structure, in which all the substituents are located on the same side of the zigzag plane representing the chain stretched out in an all-trans conformation. The structure is often represented by a "rotated Fischer projection" in which the main-chain skeleton is represented by a horizontal line (Eq. 3.1).

Isotactic polymer　　　(Rotated Fischer projection)

$$(3.1)$$

Syndiotactic polymer　　　(Rotated Fischer projection)

$$(3.2)$$

Another regular arrangement is the syndiotactic structure, in which the groups alternate from side to side and thus the configurations of the neighboring units are opposite (Eq. 3.2).

The smallest unit representing relative configuration of the consecutive monomeric units, as seen in Eq. (3.3), is termed a diad (or dyad in old references).

For a vinyl polymer, two types of diads should be considered, which are designated as meso (m) and racemo (r). Using these notations, a sequence in an isotactic polymer can be represented as –$mmmmmm$– and that in a syndiotactic polymer as –$rrrrrrrr$–. In reality, however, purely isotactic or syndiotactic polymers are rarely obtainable but the extent of regularity is always the question to be analyzed. Tacticity [1] is the term used for defining such stereochemical features of polymers.[1]

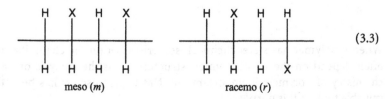

$$(3.3)$$

meso (m)   racemo (r)

NMR spectroscopy is the only analytical means that provides quantitative data on tacticity. By extending the notation of m and r, one can define relative configuration for the longer monomeric units along the chain, as triad, tetrad, pentad, and so on (n-ad in general). Equation (3.4) shows three possible triads, represented by mm, rr, and mr (the rm triad is the same as the mr triad in vinyl polymers and thus is not shown), which are also named isotactic, syndiotactic, and heterotactic triads, respectively. Similarly, for the tetrad, the following six distinguishable sequences are possible mmm, mmr, rmr, rrr, rrm, mrm, and for the pentad $mmmm$, $mmmr$, $rmmr$, $mmrm$, $mmrr$, $rmrm$, $rmrr$, $mrrm$, $mrrr$, $rrrr$.

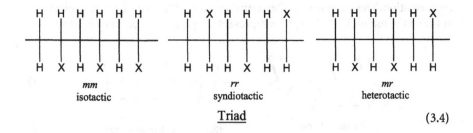

mm   rr   mr

isotactic   syndiotactic   heterotactic

<u>Triad</u>

$$(3.4)$$

3.2   Methods of Stereochemical Peak Assignments

Tacticity determination of vinyl polymers by ^{1}H NMR was first achieved for PMMA by Bovey and Tiers [2], Nishioka et al. [3] and Johnsen [4] in the beginning of the 1960s, independently. The assignments were made based on the consideration on the magnetic equivalency of the main-chain methylene protons in meso and

---

[1] The IUPAC Gold Book lists the term tacticity as "The orderliness of the succession of configurational repeating units in the main chain of a regular macromolecule, a regular oligomer molecule, a regular block or a regular chain."

racemo diads. The meso diad has no symmetry axis and H_a and H_b are nonequivalent (Eq. 3.5a). Thus H_a and H_b have different chemical shifts and exhibit AB quartet signals owing to geminal spin-coupling. Figure 3.1a shows the ^{1}H NMR spectrum of an isotactic PMMA comprising almost exclusively m diads, measured in nitrobenzene-d_5 at 110 °C, in which H_a and H_b signals are observed at 1.81 and 2.44 ppm as an AB quartet with a coupling constant of 14.6 Hz. Figure 3.1b is the spectrum of a syndiotactic PMMA, showing a singlet methylene proton signal at 2.10 ppm, which indicates that the methylene protons in the racemo diad are magnetically equivalent (Eq. 3.5b). This ^{1}H NMR spectroscopic evidence is the absolute measure of the stereochemical configuration of the vinyl polymer.

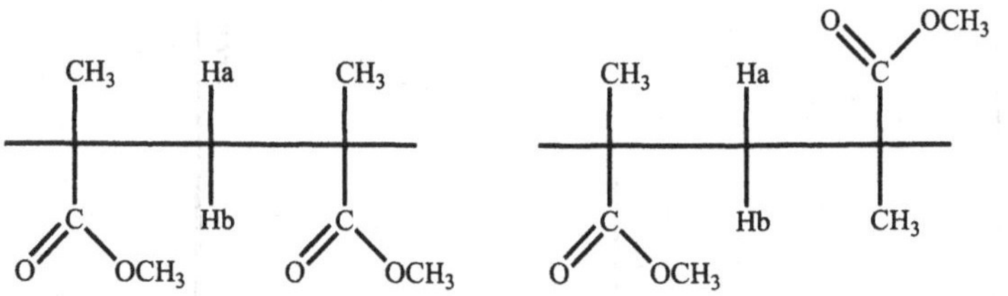

(3.5)

(A) *meso (m)*   (B) *racemo (r)*

The α-methyl proton signals in these two spectra have different chemical shifts: isotactic PMMA at 1.46 ppm (Fig. 3.1a) and syndiotactic PMMA at 1.23 ppm (Fig. 3.1b). This indicates that the α-methyl resonance is sensitive to the configuration. The smallest configurational sequence reflected in the shift of the α-methyl resonance is the triad (Eq. 3.6). The heterotactic triad signal is scarcely observed in Fig. 3.1a and b. Figure 3.1c is the spectrum of PMMA obtained by radical polymerization which comprises both *m* and *r* diads and, as a consequence, *mm*, *rr*, and *mr* triads.

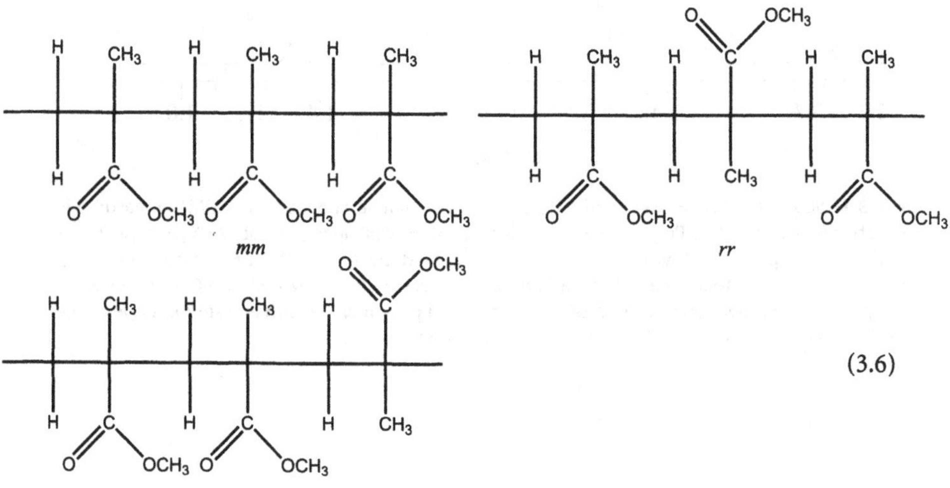

mm   *rr*

(3.6)

mr

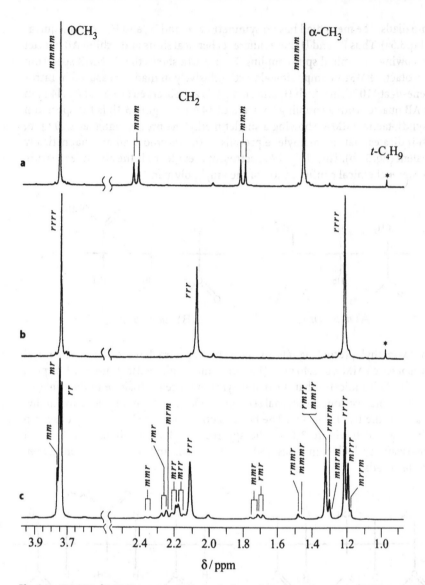

Fig. 3.1. 500 MHz ^{1}H NMR spectra of **a** isotactic, **b** syndiotactic, and **c** atactic PMMAs measured in nitrobenzene-d_5 at 110 °C. Polymer A was prepared by t-C$_4$H$_9$MgBr in toluene at –78 °C, polymer B was derived from poly(trimethylsilyl methacrylate) prepared by t-C$_4$H$_9$Li/bis(2,6-di-t-butylphenoxy)-methylaluminum in toluene at –95 °C, and polymer C was prepared by AIBN at 60 °C. 10 wt/vol%, 45° pulse, pulse repetition time 10 s, 64 scans. The small peak marked with an *asterisk* is due to the t-C$_4$H$_9$– group at the chain end introduced in the initiation step

Three groups of α-CH$_3$ signals observed in Fig. 3.1c (1.1–1.5 ppm) are thus assigned to *mm*, *mr*, and *rr* triads, respectively, from lower magnetic field. Closer inspection of the α-CH$_3$ signals reveals that each signal shows further splittings due to longer stereochemical sequences, i.e., pentads.

The methylene protons signals in Fig. 3.1c are much more complicated compared with those of Fig. 3.1a and b, reflecting the splittings due to tetrad tacticity. ^{13}C NMR signals of the carbonyl carbons of the same PMMA split mostly due to pentad tacticity, and partly due to heptad tacticity. The heptad splitting is much clear in the spectrum measured in toluene-d_8 compared with that in chloroform-d (Sect. 1.4.1, Fig. 1.16). The assignments can be made on the basis of the assumption that the stereochemical sequence distribution in radically prepared PMMA obeys Bernoullian statistics, i.e., the probability of finding the m diad at a given diad within the chain does not depend on the diad tacticity of its neighboring diad. This assumption is, strictly speaking, not correct (Sect. 3.3) but is adaptable for the peak assignment. The probability of finding the m diad, P_m, can be calculated from the diad or triad tacticity determined by ^{1}H NMR analysis. In this particular PMMA, P_m is 0.214. On the basis of this value, the fractions of longer tactic sequences can be calculated as shown in Table 3.1. In this procedure, some of pentads should have the same probability, such as *mmrm* and *mmmr*, *mmrr* and *rmrm*, and *rmrr* and *rrrm*. As seen in Fig. 3.1c, however, the peaks may be apparently grouped in three, which are *mm*-, *mr*-, and *rr*-centered sequences, and thus the pair of *mmrr* and *mrmr* is interchangeable but the pairs of *mmrm* and *mmmr* and of *rmrr* and *rrrm* are not interchangeable.

Table 3.1. Comparison of observed and calculated values of pentad tacticity for radically prepared PMMA

Pentad	Observed[a]	Calculated[b]
mmmm	0.00	0.00
mmmr	0.01	0.01
rmmr	0.03	0.03
mmrm	0.01	0.01
mmrr	0.05	0.06
rmrm	0.06	0.06
rmrr	0.22	0.21
mrrm	0.03	0.03
rrrm	0.19	0.21
rrrr	0.40	0.38

[a] Determined from the carbonyl carbon signals in the spectrum measured in toluene-d_8.

[b] The values were calculated from the probability P_m (=0.214) assuming that the stereochemical sequence distribution in radically prepared PMMA obeys a Bernoullian statistics. P_m was calculated from the triad tacticity obtained from the α-CH$_3$ proton signals.

Chemical transformation of the polymer may serve as a useful method for tacticity determination. Since ^{1}H NMR spectroscopic determination of tacticity was established for PMMA, many polymethacrylates have been converted to PMMA through hydrolysis and subsequent methylation with diazomethane to be analyzed for tacticity [5]. Nowadays, ^{13}C NMR has become very common and direct tacticity analysis of polymethacrylates has been made by ^{13}C NMR spectroscopy, in which the carbonyl carbon signals show similar tacticity splittings as that of PMMA (Sect. 1.4.1, Fig. 1.16). However, the carbonyl carbon signal of poly(2,4,6-trichlorophenyl methacrylate) does not show a splitting due to tacticity [6]. In such a case, the derivation of the polymer is still an important method.

Poly(vinyl acetate) is usually converted to poly(vinyl alcohol) for tacticity determination; assignments of ^{1}H NMR and ^{13}C NMR spectra of the latter have been well established [7, 8]. The ^{13}C NMR of poly(vinyl acetate) does show splittings due to tacticity but the peak assignment is not simple enough [9, 10].

In the case of vinyl polymers, spin-coupling between CH and CH_2 protons makes their ^{1}H NMR spectra complicated. In such cases, deuteration of specific protons in the monomer has been employed [1]. Once the tacticity of a polymer obtained under specified polymerization conditions is established by using the deuterated monomer, ^{13}C NMR may be more convenient for routine analysis.

When the overlapping signals arising from side-chain units, whose T_1 are larger than those of the main-chain signals, interfere with the tacticity analysis, a peak-elimination method is useful [11]. In the ^{1}H NMR spectrum of polymers of alkyl methacrylates, the resonance of ester groups often overlaps with α-methyl signals which are the source of tacticity data. By capitalizing on the relatively large difference in ^{1}H T_1 values of the ester group and the α-methyl group, one can eliminate the ester group signals overlapped with the α-methyl signals by employing an inversion-recovery pulse sequence, $180°-t-90°-T$; the value of t is adjusted such that the longitudinal magnetization of the ester group protons is zero.

An example of the method is demonstrated in Fig. 3.2 for poly(ethyl methacrylate) (PEMA). In the ^{1}H NMR spectrum measured in $CDCl_3$, the α-methyl signal is obscured by the overlap with the signal of methyl protons in the ester ethyl group (Fig. 3.2a). When the spectrum was taken using a pulse sequence of ($180°-0.8$ s$-90°-20$ s), the ester methyl signal was eliminated and the three splittings in the α-methyl resonance clearly appeared as shown in Fig. 3.2b. The triad fractions thus obtained agreed well with those of the PMMA derived from the PEMA. The diad tacticity calculated from the triad tacticity of the PEMA is consistent with the diad tacticity observed from methylene proton signals. The magnetization of α-methyl protons did not recover completely under the condition $t=0.8$ s, but the accurate determination of tacticity is possible owing to the relatively small difference among the T_1 values of the α-methyl protons in the three different triads.

An NMR shift reagent has been successfully used to attain better splittings due to tacticity for polymers having polar groups comprising hetero atoms such as poly(alkyl vinyl ether)s (Sect. 1.4.4, Fig. 1.23).

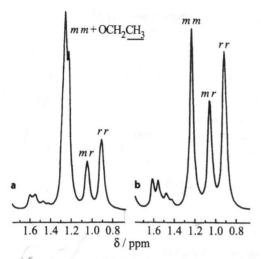

Fig. 3.2a,b. 270 MHz ^{1}H NMR spectra of PEMA in CDCl$_3$ at 55 °C [11]. **a** Normal spectrum, **b** spectrum obtained with a pulse sequence of (180°–0.8 s – 90°–20 s). 90° pulse, pulse repetition time 20 s, 32 scans

When highly stereoregular polymers are available which can be epimerized by any chemical reactions, the epimerization of the polymers provides the opportunity for tactic sequence assignments. The epimerization introduces irregular tactic sequence within the stereoregular polymer chain and so newly observed NMR signals after epimerization can be assigned. The successful examples include polypropylene [12, 13] and polystyrene [14–16].

NMR chemical shift prediction by quantum chemical calculation and the γ-gauche effect method has become a useful approach for stereochemical assignment of vinyl polymers. For example, ^{13}C NMR assignments for a series of polyolefins such as poly(1-pentene) and poly(1-octene) were demonstrated successfully [17].

To confirm the stereochemical assignments, it is advisable to check the necessary relationship among the probabilities of occurrence of the various stereosequences observed [18], which is entirely independent of the statistics of the polymerization processes (Table 3.2).

Two-dimensional NMR (Chap. 6) has been used for the assignments of stereochemically sensitive peaks, i.e., tacticity assignments. An early successful example is ^{1}H correlation spectroscopy (COSY) analysis of poly(vinyl alcohol) [19, 20]. The principle for the assignment is correlations between diad and triad, or between triad and tetrad. Consider a vinyl polymer, $+CH_2 - CHX+_n$, and assume CH_2 protons show diad splittings (m and r) and CH proton triad splittings (mm, mr, and rr). The CH_2 proton signals of the m diad may have two chemical shifts owing to the nonequivalency of the two methylene protons and should have correlation with mm and mr triads, since these two triad comprise the m diad. The signals due to the r diad may be a singlet and should have correlation with rr and mr triads. Thus among the three triad signals, the mr peak should have correlations with both m and r diad peaks. The mm peak should have two correlation peaks

Table 3.2. Some necessary relationships among sequence frequencies [18]

Diad	$(m)+(r)=1$
Triad	$(mm)+(mr)+(rr)=1$
Diad–triad	$(m)=(mm)+1/2(mr)$
	$(r)=(rr)+1/2(mr)$
Triad–tetrad	$(mm)=(mmm)+1/2(mmr)$
	$(mr)=(mmr)+2(rmr)=(mrr)+2(mrm)$
	$(rr)=(rrr)+1/2(mrr)$
Tetrad–tetrad	Sum=1
	$(mmr)+2(rmr)=2(mrm)+(mrr)$
Pentad–pentad	Sum=1
	$(mmmr)+2(rmmr)=(mmrm)+(mmrr)$
	$(mrrr)+2(mrrm)=(rrmr)+(rrmm)$
Tetrad–pentad	$(mmm)=(mmmm)+1/2(mmmr)$
	$(mmr)=(mmmr)+2(rmmr)=(mmrm)+(mmrr)$
	$(rmr)=1/2(mrmr)+1/2(rmrr)$
	$(mrm)=1/2\ (mrmr)+1/2\ (mmrm)$
	$(rrm)=2(mrrm)+(mrrr)=(mmrr)+(rmrr)$
	$(rrr)=(rrrr)+1/2(mrrr)$

with the two methylene signals of the m diad and the rr peak should have one correlation peak with the singlet r diad signal (Fig. 3.3). These lead to the peak assignments of the splittings due to diad and triad tacticities.

In the actual COSY spectrum (Sect. 6.2) of poly(vinyl alcohol), the peaks are assigned in terms of triad–tetrad correlations as follows:

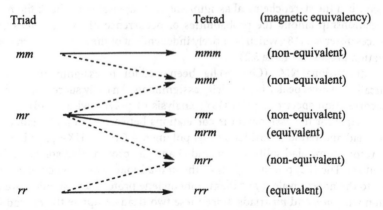

Triad		Tetrad	(magnetic equivalency)
mm		mmm	(non-equivalent)
		mmr	(non-equivalent)
mr		rmr	(non-equivalent)
		mrm	(equivalent)
		mrr	(non-equivalent)
rr		rrr	(equivalent)

(the solid arrows represent unequivocal connectivity, and the connectivity shown by the broken arrows indicates that the tetrad may show cross-peaks with two triads).

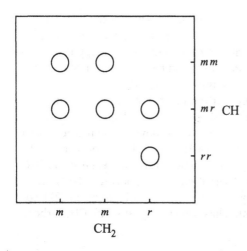

Fig. 3.3. Schematic diagram of 2D NMR spectrum of a vinyl polymer for tacticity assignment

The assignments shown in Fig. 3.4 are made on the basis of the previously described connectivity and on consideration of the magnetic equivalency of the methylene protons. The triad showing the largest number of cross peaks can be assigned to *mr*. Thus the other two are either an *mm* or an *rr* triad. Among the tetrads, *mmm* and *rrr* show a single cross-peak with an *mm* or an *rr* triad peak. The cross-peaks designated a, b, and c are the candidates. The methylene protons in the *mmm* tetrad are nonequivalent and exhibit two chemical shifts, and thus cross-peaks b and c should be assigned to *mmm-mm* correlation. Cross-peak a is then assigned to *rrr–rr* correlation. The *mrr* tetrad peaks show two cross-peaks with *rr* and *mr* triads. The *mmr* tetrad methylene peaks shows cross-peaks with *mm* and *mr* triads. The existence of only a single peak for mmr methylene protons indicates that the two methylene protons are accidentally equivalent. Both *mrm* and *rmr*

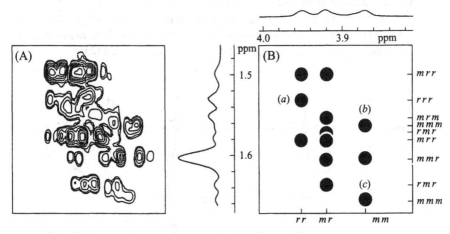

Fig. 3.4a,b. 500 MHz broadband decoupled COSY spectra of poly(vinyl alcohol) in D₂O at 80 °C [20]. **a** Expansion of CH–CH₂ cross-peaks and diagonal peaks of CH. **b** Schematic representation of **a**

tetrads show single cross-peaks with the mr triad. However, since the mrm methylene protons are magnetically equivalent and the rmr ones are nonequivalent, these two can be differentiated by using 2D J-resolved spectroscopy (Sect. 6.8) which demonstrates the spin-coupling for rmr tetrad peaks. The ^{1}H NMR assignments have been extended to ^{13}C NMR assignments using ^{13}C–^{1}H COSY (Sect. 6.3).

As described previously, the carbonyl carbon of PMMA shows splittings due to pentad tacticity. Thus the heteronuclear multiple-band correlation (HMBC) spectrum (Sect. 6.4) of PMMA enriched with ^{13}C at the carbonyl carbon has been used to provide unambiguous assignments of CH_2, α-CH_3, and OCH_3 proton signals [21].

A 2D incredible natural abundance double quantum transfer experiment (INADEQUATE) (Sect. 6.7) enables direct observation of neighboring ^{13}C–^{13}C correlation and has been applied to poly(vinyl alcohol) [20] and polypropylene [22]. For example, the CH carbon in the *rrrr* pentad has two cross-peaks with CH_2 carbons in *rrrrr* and *rrrrm* hexads (Eq. 3.7).

$$(3.7)$$

In ^{19}F NMR, long-range spin-coupling is usually important. For poly(vinyl fluoride) stereochemical assignments have also been made at the pentad level from the ^{19}F COSY spectrum, in which the cross-peaks arise from four-bond spin-coupling ($J_{F-F} \cong 7$ Hz) (F*–C–C–C–F*) between the central part of the fluorines in pentad sequences such as *rmmr* and *mmrm* that share a common tetrad [23].

$$(3.8)$$

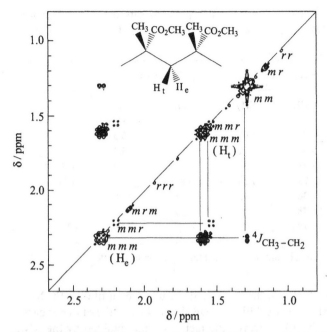

Fig. 3.5. 500 MHz ¹H COSY spectrum of isotactic PMMA measured in 10 wt/vol% solution in chlorobenzene-d_5 at 100 °C [24]

¹H COSY of isotactic PMMA (Fig. 3.5) enables unambiguous assignment of erythro (H_e) and threo (H_t) methylene protons based on the fact that only the erythro proton is capable of forming with α-methyl protons a "W-shaped" four-bond path of long-range coupling (W rule) and shows the cross-peak with α-methyl proton in *mm* triads (Sect. 6.2) [24].

3.3 Quantitative Determination of Tacticity

Since NMR spectroscopy is the only means to quantify tacticity data, the accuracy and the precision of the intensity measurement are of prime importance. In the tacticity measurement, the relative intensity of split signals of the same type of protons or carbons is to be considered. Thus the conditions suitable for quantitative measurements are not the same as those discussed in Chap. 2. The effect of pulse repetition time on the tacticity determination of a PMMA prepared by radical polymerization is shown in Table 3.3 together with the peak intensities of methoxy and methylene protons relative to that of α-methyl signals. Though the quantitative measurements for the latter require repetition times exceeding 4–6 s, the tacticity values are almost constant at a much shorter repetition time of 1.0 s. In the case of ¹³C NMR analysis of PMMA, three different sorts of carbons, α-CH₃, quaternary, and carbonyl carbons, give the tacticity data. The data obtained from these three

Table 3.3. Effect of pulse repetition time on relative peak intensity and observed tacticity for radically prepared PMMA by 500 MHz ^{1}H NMR. Solvent CDCl$_3$, 10 wt/vol%, 55 °C; pulse width 90°, number of scans 4–16

Pulse repetition (s)	Relative intensity[a]		Tacticity (%)		
	OCH$_3$	CH$_2$	*mm*	*mr*	*rr*
15.0	3.02	1.99	3.7	35.3	61.0
10.0	2.99	2.01	3.8	35.2	61.0
6.0	2.91	2.01	3.8	35.2	61.0
4.0	2.79	1.99	3.7	35.0	61.3
2.0	2.36	1.98	3.8	35.0	61.2
1.0	1.78	1.88	3.7	34.9	61.4
0.3	1.18	1.62	3.3	34.4	62.3

[a] Relative intensity against α-CH$_3$ signals whose intensity is set to be 3.00.

signals (125 MHz) are shown in Table 3.4 [25]. In usual quantitative ^{13}C NMR measurements, the peak intensity without NOE should be measured under gated decoupling conditions (NNE). However, the tacticity data obtained by the NNE mode and the complete decoupling condition with NOE (COM mode) are consistent with each other. This is due to the small difference in the NOE values for the peaks due to different tactic sequences. Thus the tacticity determination by ^{13}C NMR can be made under complete decoupling conditions, which provide a better S/N ratio with the effect of NOE.

Signals from end groups of a polymer, when they overlap with the signals of the main-chain units used for tacticity determination, might interfere with the accurate estimation of tacticity, and are sometimes not negligible as in the case of relatively low molecular weight polymers. Figure 3.6 illustrates 500 MHz and 100 MHz ^{1}H NMR spectra of a PMMA prepared with azobis(isobutyronitrile) (AIBN) [26]. In the 500 MHz spectrum two methyl proton signals due to the initiator fragment, 1-cyano-1-methylethyl group, are observed at the valley of the mm and mr triad signals (Eq. 3.9). Thus the triad tacticity data could be obtained after correcting them for the contribution of the AIBN fragment signals. In the 100 MHz spectrum the signals cannot be distinguished and thus the mm and mr triads should be overestimated. The fractions of isotactic and heterotactic triads obtained from 100 MHz ^{1}H NMR are slightly but meaningfully larger than those obtained from 500 MHz ^{1}H NMR as shown in Table 3.4. This is due to the overlap of the α-CH$_3$ proton

$$CH_3-\underset{\underset{CN}{|}}{\overset{\overset{CH_3}{|}}{C}}-CH_2-\underset{\underset{\overset{|}{C=O}}{\underset{|}{OCH_3}}}{\overset{\overset{CH_3}{|}}{C}}-CH_2-\underset{\underset{\overset{|}{C=O}}{\underset{|}{OCH_3}}}{\overset{\overset{CH_3}{|}}{C}}\text{\tiny\textasciitilde\textasciitilde\textasciitilde} \tag{3.9}$$

Table 3.4. Mean values of tacticity of PMMA determined from various NMR signals [25]

NMR signal		Tacticity (%)			$4(mm)(rr)$
		mm	mr	rr	$(mr)^2$
α-C$\underline{H}_3$	100 MHz	3.55	34.96	61.49	0.714
	100 MHz[a]	2.92	34.72	62.36	0.604
	500 MHz[a]	3.17	34.52	62.31	0.663
α-$\underline{C}H_3$	COM[b]	3.52	36.06	60.42	0.654
	NNE[c]	3.52	35.06	61.42	0.703
Quat C	COM[b]	3.10	34.31	62.59	0.659
	NNE[c]	3.34	33.69	62.97	0.741
C=O	COM[b]	3.55	34.03	62.42	0.765
	NNE[c]	3.38	34.50	62.12	0.707

[a] Corrected for AIBN fragment signal (see text).
[b] Complete decoupling condition.
[c] Gated decoupling condition without NOE.

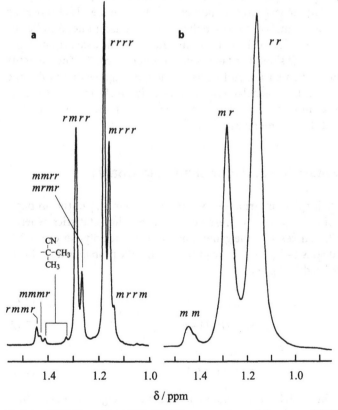

Fig. 3.6. a 500 MHz and **b** 100 MHz ^{1}H NMR spectra of α-methyl protons of PMMAs prepared with AIBN in benzene at 60 °C [26]. 45° pulse, pulse repetition time 10 s, 32 scans

signals with the initiator fragment signals. The triad tacticity from 100 MHz ^{1}H NMR corrected for the overlap agrees well with those from 500 MHz ^{1}H NMR(Table 3.4). In Table 3.4 are also shown the values of $4(mm)(rr)/(mr)^2$ that should be unity when the polymerization can be described by Bernoullian statistics. The values for the radically prepared PMMA are clearly less than unity, indicating that the statistics of the radical polymerization of MMA deviate slightly from Bernoullian [25, 26].

Isotactic PMMA prepared with t-C$_4$H$_9$MgBr has the following structure:

$$CH_3-\underset{\underset{CH_3}{|}}{\overset{\overset{CH_3}{|}}{C}}-CH_2-\underset{\underset{\underset{\underset{OCH_3}{|}}{C=O}}{|}}{\overset{\overset{CH_3}{|}}{C}}-CH_2-\underset{\underset{\underset{\underset{OCH_3}{|}}{C=O}}{|}}{\overset{\overset{CH_3}{|}}{C}}\text{\small{wwwwwww}}CH_2-\underset{\underset{\underset{\underset{OCH_3}{|}}{C=O}}{|}}{\overset{\overset{CH_3}{|}}{C}}-H \qquad (3.10)$$

The t-C$_4$H$_9$ proton signal overlaps with the rr triad signal when measured in chloroform-d but is observed separately when measured in nitrobenzene-d_5. Thus, in order to obtain correct tacticity data, it is advisable to take the NMR spectrum in nitrobenzene-d_5. An example of the spectrum is shown in Fig. 3.7. The signals from the first three and the last three monomeric units at both ends also overlap with the in-chain α-CH$_3$ signals as shown in the figure. The assignments indicated in the figure were made by ^{1}H COSY. In order to determine the exact triad tacticity, these overlapped signals should be taken into consideration [27]. The isotactity of the in-chain units, when corrected for the overlapped signals due to the end units, is independent of the molecular weight of the polymer. Without correction, the isotacticity apparently increases with molecular weight and would lead to a misunderstanding of the polymerization reaction [27].

3.4 Sequence Statistics and Propagation Mechanism

The analysis of tacticity gives information on the stereochemistry of the propagation reaction. The simplest case is represented by Bernoullian statistics, where a single probability P_m can describe the generation of the tactic polymer chain. The probability of an r diad is $1-P_m$. It can be readily seen that the probabilities of forming mm, mr, and rr triads are given by

$$[mm] = P_m^2, \qquad (3.11)$$

$$[mr] = 2P_m(1-P_m), \qquad (3.12)$$

and

$$[rr] = (1-P_m)^2. \qquad (3.13)$$

The probability of the mr triad is given double statistical weighing because both directions mr and rm must be counted.

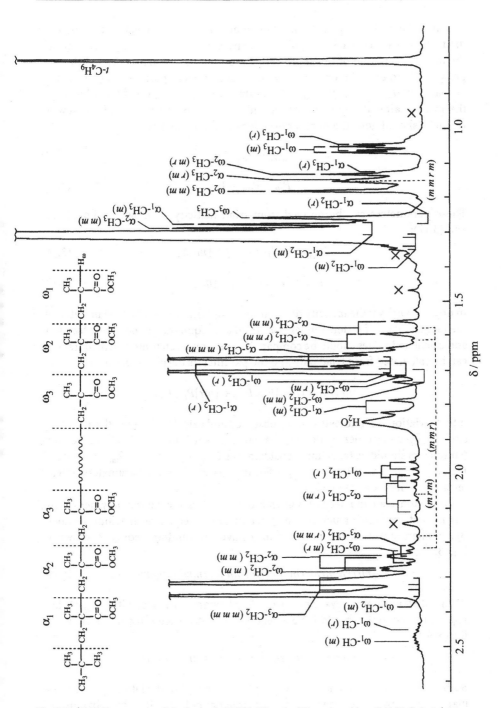

Fig. 3.7. ¹H NMR spectrum of the isotactic PMMA (15 wt/vol%) prepared by t-C₄H₉MgBr in toluene at −78 °C [27]. Signals due to the end groups and monomeric units at and near the α and ω ends are indicated according to the numbering system shown in the figure

When the probability of forming the m or the r diad depends on the previous diad, the first-order Markov sequence is generated by propagating steps in which the mode of addition of the approaching monomer is influenced by whether the growing chain end is m or r. We now have four kinds of parameters (conditional probability), $P_{m/m}$, $P_{r/m}$, $P_{r/r}$, $P_{m/r}$, for example, $P_{m/m}$ is the probability of forming the m diad after the growing chain end with the m diad. Since the following relations should hold, the number of independent parameters is 2:

$$P_{m/m} + P_{m/r} = 1, \tag{3.14}$$

$$P_{r/r} + P_{r/m} = 1. \tag{3.15}$$

These parameters can be derived from triad tacticity as shown in Eqs. (3.16) and (3.17):

$$P_{m/r} = [mr]/(2[mm] + [mr]), \tag{3.16}$$

$$P_{r/m} = [mr]/(2[rr] + [mr]). \tag{3.17}$$

In order to verify if the tacticity distribution obeys first-order Markovian statistics, the tacticity data of orders higher than triad are required, and observed fractions and calculated values should be compared. For example, an mmrm pentad fraction can be calculated from triad data and the parameters derived thereof as

$$[mmrm] = [mm]P_{m/r}P_{r/m} + [mr]P_{r/m}P_{m/m}. \tag{3.18}$$

If the deviation is found between the observed and calculated values, there is a possibility of second-order Markovian statistics, in which four kinds of independent parameters should be taken into consideration; $P_{mm/r}$ $(=1-P_{mm/m})$, $P_{mr/m}$ $(=1-P_{mr/r})$, $P_{rm/r}$ $(=1-P_{rm/m})$, and $P_{rr/m}$ $(=1-P_{rr/r})$. For the analysis of these parameters, at least tetrad tacticity data are needed.

To demonstrate the need of higher-order statistics, consider a polymer having an m diad fraction of 0.5. If the stereochemical sequence distribution obeys Bernoullian statistics, the polymer may have the random sequence shown in Eq. (3.19).

$$- - -mrrrmmrmrrmmmrrmrmmmrrmrrmmmr- - -. \tag{3.19}$$

If the distribution obeys first-order Markovian statistics with $P_{m/r}=1$ and $P_{r/m}=1$, the polymer should have a regular sequence (heterotactic) as shown in Eq. (3.20).

$$- - -mrmrmrmrmrmrmrmrmrmrmrmrmrmr- - -. \tag{3.20}$$

Both chains comprise equal number of m and r diads, but the latter is a stereoregular polymer and the former is a nonstereoregular polymer. It is thus obvious that the analysis of longer configurational sequence is important to know the higher level of stereoregularity.

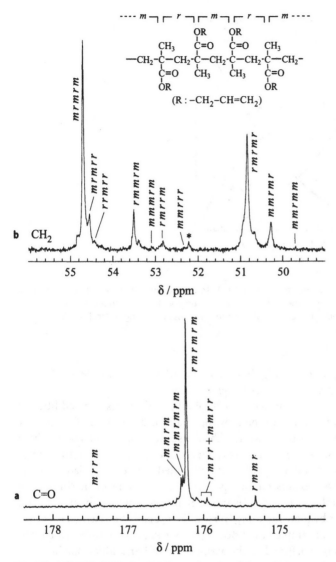

Fig. 3.8. 125 MHz NMR signals of **a** carbonyl and **b** methylene carbon signals of heterotactic poly(allyl methacrylate) (DP=88.4) measured in CDCl₃ at 55 °C (mr triad content 95.8%). The peak marked with an *asterisk* may be due to methylene carbons of terminal monomer units

Methylene and carbonyl carbon signals of a heterotactic poly(allyl methacrylate) with an mr triad content of 95.8% are shown in Fig. 3.8. In the carbonyl carbon resonance, the *mr*-centered peaks are most abundant, reflecting the high heterotacticity. In contrast, the methylene signals are composed of *m*-centered and *r*-centered peak groups with almost the same intensity. If you derive only the diad tacticity from the methylene signals, you may erroneously conclude that the polymer is nonstereoregular; however, once you obtain the triad tacticity, you will

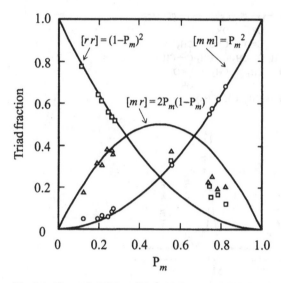

Fig. 3.9. The probabilities of isotactic (mm, *circles*), heterotactic (mr, *triangles*), and syndiotactic (rr, *squares*) triads as a function of P_m, the probability of m placement. The *points* on the left-hand side are for MMA polymer prepared with free-radical initiators and those on the right-hand side for polymer with anionic initiators [28]

soon realize that the polymer is highly stereoregular. This example clearly demonstrates the importance of the longer-sequence analysis.

When a polymer chain is formed by a Bernoulli trial process, the addition is influenced only by the end unit of the growing chain, and the probability of m or r addition (P_m or $1-P_m$) may be determined by the steric requirements of the chain end and the incoming monomer. For example, in the radical polymerization of MMA predominantly syndiotactic polymer is obtained (Table 3.1). Triad fractions [mm], [mr], and [rr] are expressed by Eqs. (3.11), (3.12), and (3.13). When triad tacticity data are available, one can examine if the polymer is formed with a Bernoulli trial process by plotting [mm], [mr], and [rr] against P_m, which is known as Bovey's plot [28]. An example of the Bovey plot for PMMAs of different tacticities [28] is shown in Fig. 3.9. Based on the assumption of Bernoullian statistics, P_m can be derived from every triad value; however, the largest value is used owing to the accuracy, i.e., [mm] is used for isotactic-rich polymers and [rr] for syndiotactic-rich ones. The plots on the left-hand side of the figure ($P_m<0.5$) fit well with the expected lines of [mm], [mr], and [rr], indicating the corresponding radically prepared PMMAs are formed by Bernoulli trial processes. In contrast, the plots on the right-hand side ($P_m>0.5$), for the PMMAs which are prepared by anionic initiators, deviate from the lines of mr and rr triads (The values of P_m were calculated from the fractions of the mm triad.), suggesting that the higher-order statistical process is required for describing the process.

In coordination polymerization, such as the polymerization of α-olefins with Ziegler catalysts, an enantiomer-selective site model has been employed for the analysis of the configurational sequence [29]. In this model, a parameter σ, the probability of forming a d (or l) unit at the d- (or l-) selective site, is used to define the tacticity:

$$[mm] = 1 - 3\sigma(1 - \sigma), \tag{3.21}$$

$$[mr] = 2\sigma(1 - \sigma), \tag{3.22}$$

$$[rr] = \sigma(1 - \sigma), \tag{3.23}$$

Thus, the mr triad fraction should be twice as large as the rr triad.

Stereochemical defects in an isotactic polymer formed according to this model can be depicted as follows:

$$
\begin{array}{c}
\quad m \qquad m \qquad m \qquad r \qquad r \qquad m \qquad m \qquad m \\
\quad\; X \qquad X \qquad X \qquad X \qquad H \qquad X \qquad X \qquad X \qquad X \\
-CH_2\text{-}\overset{|}{\underset{|}{C}}\text{-}CH_2\text{-}\overset{|}{\underset{|}{C}}\text{-}CH_2\text{-}\overset{|}{\underset{|}{C}}\text{-}CH_2\text{-}\overset{|}{\underset{|}{C}}\text{-}CH_2\text{-}\overset{|}{\underset{|}{C}}\text{-}CH_2\text{-}\overset{|}{\underset{|}{C}}\text{-}CH_2\text{-}\overset{|}{\underset{|}{C}}\text{-}CH_2\text{-}\overset{|}{\underset{|}{C}}\text{-}CH_2\text{-}\overset{|}{\underset{|}{C}}- \\
\quad\; H \qquad H \qquad H \qquad H \qquad X \qquad H \qquad H \qquad H \qquad H \\
(d) \quad (d) \quad (d) \quad (d) \quad (l) \quad (d) \quad (d) \quad (d) \quad (d)
\end{array}
\tag{3.24}
$$

in which a defect unit with the opposite configuration (l) is introduced within the sequence comprising units with d configuration. That is, the sequence does not contain an isolated r diad. From the pentad level analysis, the following relation should be kept among the pentads involving r diads:

$$[mmmr] = [mmrr] = 2[mrrm]. \tag{3.25}$$

When the stereosequence of a highly isotactic polymer is determined by the relative configuration as represented by a probability such as P_m, the defect is depicted by the following:

$$
\begin{array}{c}
\quad m \qquad m \qquad m \qquad r \qquad m \qquad m \qquad m \\
\quad\; X \qquad X \qquad X \qquad X \qquad H \qquad H \qquad H \qquad H \\
-CH_2\text{-}\overset{|}{\underset{|}{C}}\text{-}CH_2\text{-}\overset{|}{\underset{|}{C}}\text{-}CH_2\text{-}\overset{|}{\underset{|}{C}}\text{-}CH_2\text{-}\overset{|}{\underset{|}{C}}\text{-}CH_2\text{-}\overset{|}{\underset{|}{C}}\text{-}CH_2\text{-}\overset{|}{\underset{|}{C}}\text{-}CH_2\text{-}\overset{|}{\underset{|}{C}}- \\
\quad\; H \qquad H \qquad H \qquad H \qquad X \qquad X \qquad X \qquad X
\end{array}
\tag{3.26}
$$

In this case, the polymer scarcely contains an rr triad as long as the isotacticity is sufficiently high. Thus, in the analysis of highly isotactic polymers, the quantitative determination of the minor sequences is important to distinguish the polymerization mechanisms. An example of this type of isotactic polymer is a PMMA prepared with t-C_4H_9MgBr in toluene at low temperature. The carbonyl carbon NMR signals of the isotactic PMMA are displayed in Fig. 3.10, in which signals due to $mmrm$ and $mmmr$ of equal intensity are observed besides the strong $mmmm$ pentad signal [30].

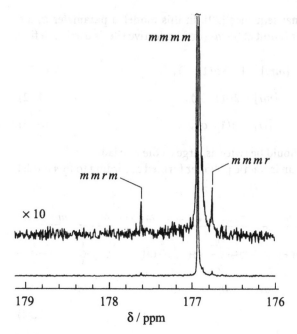

Fig. 3.10. ^{13}C NMR signal of the carbonyl carbon in the PMMA prepared by t-C$_4$H$_9$MgBr in toluene at –78 °C. Nitrobenzene-d_5, 110 °C, 100 MHz [30]

References

1. BOVEY FA (1982) Chain structure and conformation of macromolecules. Academic, New York
2. BOVEY FA, TIERS GVD (1960) J Polym Sci 44:173
3. NISHIOKA A, WATANABE H, ABE K, SONO Y (1960) J Polym Sci 48:241
4. JOHNSEN U (1962) Kolloid Z 178:161
5. KATCHALSKY A, EISENBERG H (1951) J Polym Sci 6:145
6. HILLER W-G, PASCH H, LAMPE IV (1991) Makromol Chem 192:1431
7. MORITANI T, KURUMA I, SHIBATANI K, FUJIWARA Y (1972) Macromolecules 5:577
8. OVENALL DW (1984) Macromolecules 17:1458
9. WU TK, OVENALL DW (1974) Macromolecules 7:776
10. SUNG HN, NOGGLE JH (1981) J Polym Sci Polym Phys Ed 19:1593
11. HATADA K, OHTA K, OKAMOTO Y, KITAYAMA T, UMEMURA Y, YUKI H (1976) J Polym Sci Polym Lett Ed 14:531
12. STEHLING FC, KNOX JR (1975) Macromolecules 8:595
13. SUTER UW, NEUENSCHWANDER P (1981) Macromolecules 14:528
14. YOON DY, FLORY PJ (1977) Macromolecules 10:562
15. SCHEPHERD L, CHEN TK, HARWOOD HJ (1979) Polym Bull 1:445
16. HARWOOD HJ, CHEN TK, DASGUPTA A, KINSTLEJ F, SAUFEE ER, SHEPHERD L, LIN F-T (1985) Polym Prepr Am Chem Soc Div Polym Chem 26:39
17. ASAKURA T, DEMURA M, NISHIYAMA Y (1991) Macromolecules 24:2334
18. FRISCH HL, MALLOWS CL, BOVEY FA (1966) J Chem Phys 45:1565

19. GIPPERT GP, BROWN LR (1984) Polym Bull 11:585
20. HIKICHI K, YASUDA M (1987) Polym J 19:1003
21. MOAD G, RIZZARDO E, SOLOMON DH, JOHNS SR, WILLING RI (1986) Macromolecules 19:2494
22. MIYATAKE T, KAWAI Y, SEKI Y, KAKUGO M, HIKICHI K (1989) Polym J 21:809
23. BRUCH MD, BOVEY FA, CAIS RE (1984) Macromolecules 17:2547
24. SCHILLING FC, BOVEY FA, BRUCH MD, KOZLOWSKI SA (1985) Macromolecules 18:1418
25. HATADA K, TERAWAKI Y, KITAYAMA T, UTE K (1989) Polym Prepr Jpn 38:845 (English edition E345)
26. HATADA K, KITAYAMA T, TERAWAKI Y, CHÛJÔ R (1987) Polym J 19:1127
27. HATADA K, UTE K, TANAKA K, IMANARI M, FUJII N (1987) Polym J 19:425
28. BOVEY FA (1969) Polymer conformation and configuration. Academic, New York, p 13
29. SHELDEN RA, FUENO T, TSUNETSUGU T, FURUKAWA J (1965) J Polym Sci B 3:23
30. HATADA K, UTE K, TANAKA K, KITAYAMA T (1986) Polym J 18:1037

4 Copolymer

4.1 NMR Analysis of Composition and Sequence in Copolymers

Structural analysis of copolymers is very important from both scientific and industrial points of views. High-resolution NMR analysis is particularly effective in the study of vinyl copolymers and reveals the structural and sequence details that cannot be detected by any other means. Copolymer composition can be determined very easily by NMR, particularly by ^{1}H NMR, and the method is usually more accurate than other traditional analytical methods, such as elemental analysis.

The 500 MHz ^{1}H NMR spectra of copolymers of MMA and AN [poly(MMA-co-AN)s] with three different comonomer compositions (samples 1, 2 and 3) are

Table 4.1. Compositional analysis of poly(MMA-co-AN) by 500 MHz ^{1}H NMR with a single spectrometer [1, Terawaki Y, Kitayama T, Hatada K unpublished results]. The figures in *parentheses* represent the standard deviation σ (%)

Run	MMA unit contents of the copolymer (%)		
	Sample 1	Sample 2	Sample 3
1	36.9	50.3	66.4
2	36.9	50.5	66.6
3	37.0	50.3	67.0
4	36.9	50.4	66.7
5	36.7	50.6	65.4
Average	36.9 (0.27)	50.4 (0.23)	66.4 (0.82)
Average[a]	37.5 (3.72)	52.0 (9.45)	67.1 (6.66)
^{13}C NMR[b]	36.1 (4.39)	49.9 (3.90)	65.2 (3.07)
^{13}C NMR[c]	35.6 (2.10)	49.6 (1.11)	64.2 (1.36)

[a] The results obtained by a round-robin method from 45 different spectrometers whose frequencies range from 90 to 500 MHz [1].

[b] The results obtained by a round robin method from 23 different spectrometers whose frequencies range from 22.5 to 125 MHz [1].

[c] The results obtained at 125 MHz by gated decoupling without NOE with a single spectrometer (Terawaki Y, Kitayama T, Hatada K unpublished results).

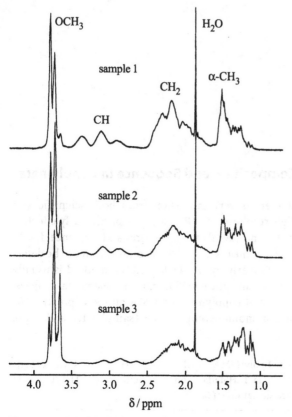

Fig. 4.1. 500 MHz ^{1}H NMR spectra of poly(MMA-*co*-AN)s in nitrobenzene-d_5 at 110 °C [1]

shown in Fig. 4.1. The copolymer compositions can be determined from the relative intensities of the signals due to the OCH$_3$ protons of MMA units and the CH proton of AN units. The results of analyses with a single spectrometer are shown in Table 4.1 (Terawaki Y, Kitayama T, Hatada K unpublished results). The standard deviations, σ, of the determinations are about 1% or less.

The series of experiments were also carried out with a round robin method by the Research Group on NMR, SPSJ [1]. The averaged results of the composition determinations with 45 different spectrometers operated at different frequencies (90–500 MHz) are also shown in Table 4.1 and agreed well with those obtained using a single 500 MHz spectrometer. The values of σ for the composition determinations by the round-robin method were about 4–9%. The deviation for sample 1 was smaller than the deviations for samples 2 and 3, which have smaller AN contents than sample 1. Larger σ values for samples 2 and 3 may be due to the fact that the methine proton signals are broad multiplets [1]. The copolymer compositions could also be determined from the 125 MHz NMR signal intensities of CO (MMA) and CN (AN) carbons with complete decoupling and agreed well with those obtained from ^{1}H NMR spectra (Table 4.1) [1, Terawaki Y, Kitayama T, Hatada K

Table 4.2. Comonomer sequences in copolymers poly(A-*co*-B) made of nonprochiral monomers [2]

Diad	AA	AB or BA	BB
Triad	AAA BAA (or AAB) BAB		BBB ABB (or BBA) ABA
Tetrad	AAAA BAAA (or AAAB) BAAB	AABA (or ABAA) BABA (or ABAB) AABB (or BBAA) BABB (or BBAB)	BBBB ABBB (or BBBA) ABBA

unpublished results]. The agreement may be ascribed to the similar NOE values for CO (1.13) and CN (1.06) carbons. ^{13}C NMR analysis with gated decoupling by a single spectrometer gave highly precise copolymer compositions that are consistent with the results from ^{1}H NMR (Table 4.1).

The properties of a copolymer depend not only on its composition but also on comonomer sequence and stereochemical sequence. Although compositional analysis can be achieved by several methods other than NMR, quantitative analysis on sequence distribution can be made only by NMR spectroscopy. If a copolymer is produced only from nonprochiral monomers, such as comonomers of the type $CH_2=CX_2$, there exists a set of sequences shown in Table 4.2; there are no stereo-

Table 4.3. Configurational sequences in copolymers poly(A-*co*-B) made of prochiral monomers

		AA	AB (or BA)	BB
diad	*m*	X X	X Y	Y Y
	r	X / X (X)	X / Y (Y)	Y / Y (Y)
		AAA	BAA (or AAB)	BAB
triad[a]	*mm*	X X X	Y X X	Y X Y
	mr	X X / X	Y X / X X X / Y	Y X / Y
	rr	X / X (X)	Y / X (X)	Y / Y (X)

[a] Ten others with BBB, ABB (or BBA) and ABA triads.

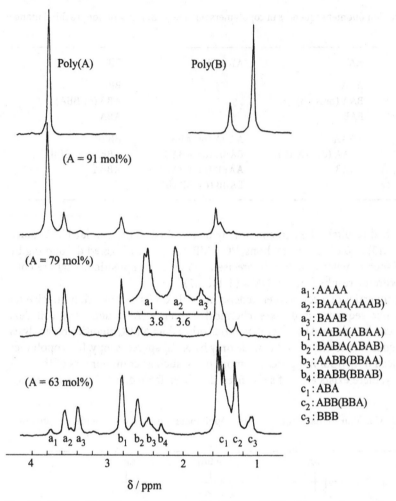

Fig. 4.2. 60 MHz ^{1}H NMR spectra measured in S_2Cl_2 at 130 °C of homopolymers of vinylidene chloride (*A*) and isobutylene (*B*), and copolymers of A and B [3]

chemical sequences [2]. If a copolymer is prepared from prochiral monomers, such as comonomers of the type CH_2=CHX or CH_2=CXY, configurational sequences are to be considered (Sect. 3.1) and there exist sets of sequences for diads and triads as shown in Table 4.3 [2]. The number of the sequences in *n*-ads increases largely with *n*.

A typical example of a copolymer that shows peak splittings only due to the comonomer sequences shown in Table 4.2 is the copolymer of vinylidene chloride (A) and isobutylene (B). The ^{1}H NMR spectra of poly(A), poly(B), and poly-(A–*co*–B)s with different compositions are shown in Fig. 4.2 [3]. Although the spectra were taken at 60 MHz many years ago, they clearly show the peak splittings due to the triad and tetrad comonomer sequences. It is natural that the spectra show no splittings due to the configurational sequences and spin–spin coupling. As a result, the peak assignments are easily and reasonably made by the inspection of the spectra of copolymers with different compositions. Three peaks in the range 3.2–3.9 ppm (a_1, a_2, a_3) can be assigned to the methylene protons ($-CCl_2-CH_2-CCl_2-$) in the AA-centered tetrads and four peaks at 2.2–3.0 ppm (b_1, b_2, b_3, b_4) methylene protons [$-CCl_2-CH_2-C(CH_3)_2-$ or $-C(CH_3)_2-CH_2-CCl_2-$] in the AB- or BA-centered tetrads. The peaks at 1.0–1.6 ppm (c_1, c_2, c_3) are overlaps of the methyl proton peaks in B-centered triads and the methylene proton peaks in the BB-centered tetrads. Although the peak separation is not enough to distinguish the triad and tetrad peaks, the whole pattern appears to show the triad distribution. All the assignments are indicated in Fig. 4.2. The fine splittings observed at peaks a_1–a_3 and c_1–c_3 may be attributed to the hexad and pentad sequences, respectively.

Even with advanced NMR methods, structural analysis of copolymers is a very difficult task when the chemical shifts of the signals are sensitive to both configurational and monomeric sequences. It should be easier to analyze signals of the copolymer with high stereoregularity since the signals have to be simple enough in the absence of splitting due to the configurational sequences. The carbonyl carbon NMR signals of isotactic, syndiotactic, and atactic copolymers of MMA and n-butyl methacrylate (n-BuMA) [poly(MMA–co–n-BuMA)] measured in chlorobenzene-d_5 at 115 °C and 125 MHz are shown in Fig. 4.3 [4]. The isotactic copolymer was prepared with t-C_4H_9MgBr in toluene and was almost completely isotactic. So the five peaks around 176.4 ppm should be ascribed to MMA (M) and n-BuMA (B) centered monomer sequences of triads in $mmmm$ configurational sequences; these are the overlap of the two triplets centered at 176.6 and 176.3 ppm due to the M- and B-centered triads as shown in Fig. 4.3a. The assignments were made by comparing the spectral patterns of the copolymers of different copolymer compositions. The intensity measurement of the five peaks indicated that the relative intensities of the three peaks in each triplet are almost 1:2:1, i.e., the copolymer is random in its monomer sequence.

The tacticity of the syndiotactic polymer that was prepared with t-C_4H_9Li/(C_2H_5)$_3$Al in toluene is rather low (86% in the rr triad) compared with that of the isotactic polymer (97% in the mm triad) and shows the signals due to mrrr and rmrr configurational sequences as well as those due to the $rrrr$ sequences in the spectrum (Fig. 4.3b). The two strong triplets centered at 177.5 and 177.2 ppm were assigned to M- and B-centered triads in $rrrr$ configurational sequences, respectively, as shown in the figure. The peak assignments were made similarly to that for the isotactic polymer. The intensity measurement of the six peaks indicated that the monomer sequence distribution in the copolymer is also random [4]. The multiplets centered at 177.8 and 176.7 ppm are due to the carbonyl carbons in $mrrr$ and $rmrr$ configurational sequences, respectively. The tactic sequence assignments are based on those for PMMA and other polymethacrylates. Fine splittings in these multiplets are due to the monomer sequence distribution. Since these configurational pentad sequences are unsymmetrical, four peaks are possible for M- and B-centered triads, respectively, as indicated in Fig. 4.3b. For example,

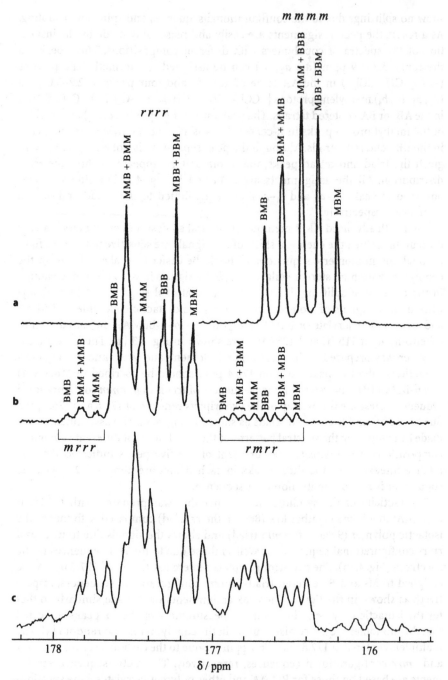

Fig. 4.3. Carbonyl carbon NMR signals of **a** isotactic (M:B=49:51), **b** syndiotactic (M:B=50:50), and **c** atactic (M:B=59:41) copolymers of MMA and *n*-BuMA prepared with *t*-C$_4$H$_9$MgBr, *t*-C$_4$H$_9$Li/(C$_2$H$_5$)$_3$Al, and AIBN, respectively, measured in chlorobenzene-d_5 at 115 °C [4]. *M* and *B* represent MMA and *n*-BuMA units in the copolymer, respectively

the M-centered triad with the rmrr configuration consists of $rBmMrBr, rMmMrBr$, $rBmMrMr$, and $rMmMrMr$ with increasing order of magnetic field.[1] The signals due to $rMmMrBr$ and $rBmMrMr$ sequences show different chemical shifts from each other, although the assignments for these two peaks cannot be made explicitly [4].

The signals of atactic poly(MMA–co–n–BuMA) (59 mol% MMA units) prepared with AIBN in toluene are shown in Fig. 4.3c. The spectrum is a complex multiplet owing to the coexistence of most of the monomeric and configurational sequences in detectable amounts. The peak assignments for the copolymer were made on the basis of the previously mentioned assignments for the isotactic and syndiotactic copolymers referring to the assignments for the stereoregular homopolymers of methacrylates and are shown in Fig. 4.4. The peak assignments were confirmed by the calculation of the fractions for all the triad monomer sequences assuming a terminal model for the copolymerization and Bernoullian statistics for stereo-regulation in the radical copolymerization [4].[2]

NMR is also useful for the structural analysis of block copolymers which are often prepared by living polymerization methods. Polymerization of methacrylates by t-C$_4$H$_9$MgBr is a living one and can produce block copolymers of methacrylates. The 500 MHz ^{1}H NMR spectra of PMMA–block–PEMA (B) and PEMA– block– PMMA (C) are shown in Fig. 4.5 together with the spectra of a mixture of PMMA and PEMA (D) and poly(MMA–co–EMA) (A) all of which were prepared using t-C$_4$H$_9$MgBr. The PMMA–block–PEMA was prepared by polymerizing EMA with PMMA anion and the PEMA–block–PMMA was prepared by polymerizing MMA with PEMA anion [5]. In the spectra of the block copolymers the methylene protons showed two sets of AB quartet signals due to the PMMA block and the PEMA block (Fig. 4.5b,c), clearly indicating that both blocks are highly isotactic. The corresponding signals of the methylene protons in the random copolymer (Fig. 4.5a) showed much more complicated splittings (1.75 ppm) or broadening

---

[1] For example, $rBmMrBr$ represents the following sequence, where X is –COOCH$_3$, Y is –COOC$_4$H$_9$, and Z is X or Y.

$$-CH_2 \cdot \underset{\underset{CH_3}{|}}{\overset{\overset{Z}{|}}{C}} - CH_2 \cdot \underset{\underset{Y}{|}}{\overset{\overset{CH_3}{|}}{C}} - CH_2 \cdot \underset{\underset{X}{|}}{\overset{\overset{CH_3}{|}}{C}} - CH_2 \cdot \underset{\underset{CH_3}{|}}{\overset{\overset{Y}{|}}{C}} - CH_2 \cdot \underset{\underset{Z}{|}}{\overset{\overset{CH_3}{|}}{C}} -$$

[2] Under the assumption of the terminal model, the monomer sequence distribution can be estimated from monomer reactivity ratios, r_M and r_B, by introducing probability parameters, $P_{MB} = 1/(1 + r_M[M]/[B])$ and $P_{BM} = 1/(1 + r_B[B]/[M])$, where M and B denote MMA and n-BuMA. For example, the probability or fraction of MMM triad, P_{MMM}, can be calculated as $P_{MMM} = P_M(1 - P_{BM})^2$ (P_M is the MMA content in the copolymer). As to the tacticity, the probability of meso addition, P_m, in the radical copolymerization was deteremined from average triad tacticity to be 0.197, in this particular case. By assuming Bernoullian statistics for stereoregulation, the fraction of MMM triad in the rrrr pentad, for example, can be calculated as $P_{MMM}(1 - P_m)^4$.

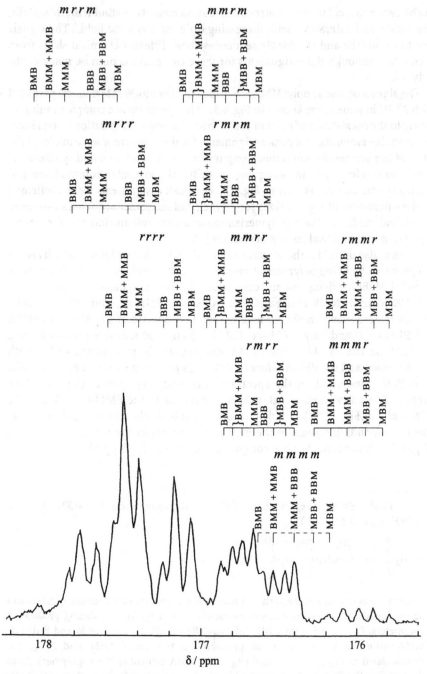

Fig. 4.4. Assignments of carbonyl carbon NMR signals of poly(MMA–*co*–*n*-BuMA) (M:B=59:41) (copolymer c in Fig. 4.3) at the triad level in the monomer sequence and at the pentad level in the configurational sequence [4]. *M* and *B* in the figure represent MMA and *n*-BuMA units in the copolymer, respectively

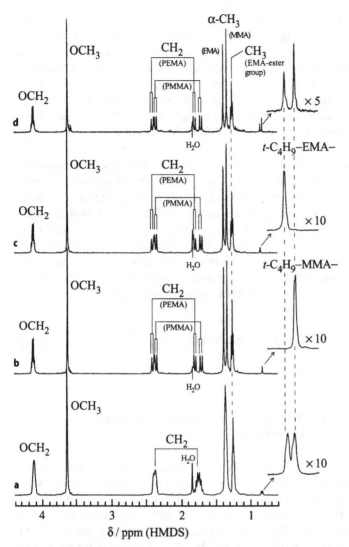

Fig. 4.5. a–d 500 MHz ^{1}H NMR spectra of highly isotactic copolymers of MMA and EMA prepared with t-C$_4$H$_9$MgBr in toluene at –60 °C, measured in nitrobenzene-d_5 at 110 °C. **a** Poly(MMA-co-EMA); **b** PMMA-$block$-poly(EMA); **c** poly(EMA)-$block$-PMMA; **d** PMMA+poly(EMA) [5]. 45° pulse, pulse repetition time 20 s, 3,000 scans

(2.4 ppm) owing to the presence of different types of monomer sequences. Similar spectral differences between the block and random copolymers are also observed in the other signals. Thus, these ^{1}H NMR signals are good and clear indications for distinguishing the block and random copolymers of MMA and EMA.

NMR spectra of the block copolymers are expected to be almost the super-positions of the spectra of the corresponding homopolymers. However, the signals due to initiator fragments (0.8–0.9 ppm) provide information for distinguishing the block copolymers from the mixture. The PMMA-*block*-PEMA (Fig. 4.5b) and PEMA-*block*-PMMA (Fig. 4.5c) showed single t-C_4H_9 signals at 0.823 and 0.852 ppm, respectively, which are assigned to the initiator fragments attached to the PMMA and PEMA sequences, respectively. On the other hand, the mixture of PMMA and PEMA, both of which were prepared with the same initiator t-C_4H_9MgBr, shows signals of two types of initiator fragments as indicated in Fig. 4.5d. The t-C_4H_9 signals in the two block copolymers provide the information whether the block copolymer was prepared from the PMMA anion or the PEMA anion [5].

In the carbonyl carbon region of the 125 MHz ^{13}C NMR spectrum measured in CDCl$_3$ at 55 °C the diblock copolymer PMMA-*block*-PEMA shows two strong signals at 176.51 and 176.37 ppm, with weak signals of equal intensity at 176.60 and 176.28 ppm. The two strong signals were assigned to the MMM and EEE triads in the isotactic pentad configuration, *mmmm*, respectively, by referring to the spectra of isotactic PMMA and isotactic PEMA. The two weak signals of equal intensity were assigned to MME and MEE triads in the *mmmm* configuration by comparing the spectrum with that of the corresponding random copolymer, which should exist at the switching point of MMA and EMA sequences:

$$\overline{}\text{MMMMMMMME}\overline{\underline{}}\text{EEEEEE}\overline{} \qquad (4.1)$$

4.2 Chemical Shift Calculation for Peak Assignment of Hydrocarbon Copolymers

For the peak assignments in ^{13}C NMR spectra of hydrocarbon polymers, particularly polypropylene and poly(ethylene-*co*-propylene), empirical additive shift rules [6–8] are sometimes very useful. Cheng and Bennett [9] devised detailed ^{13}C NMR chemical shift rules for ethylene and propylene polymers and for low-molecular-weight analogues that account for configurational sequences as well as for substituent effects. According to this rule, the shift equation for methylene carbons is given as Eq. (4.2)

$$\delta_{CH_2} = 29.9 + \sum_i A_i + \sum_{ij} B_{ij} + \sum_{ij} C_{ij} + \sum_k D_k, \qquad (4.2)$$

where A_i is the methyl-substituent parameter for position i, B_{ij} is the correction term corresponding to the configurations of the methyl groups at i and j on the same side of the carbon in question (e.g., C^β and C^δ in Eq. 4.3a), and C_{ij} is to be used for methyl groups at positions i and j on the opposite sides of the carbon in question (e.g., C^β and $C^{\beta'}$ in Eq. 4.3a). The term D_k is specifically designed to take into account the effect of three consecutive propylene units, as in Eq. (4.3), where C^* represents the carbon in question.

$$
\begin{array}{ccccccc}
C^\delta & C^\beta & C^{\beta'} & & C^\zeta & C^\delta & C^\beta \\
| & | & | & & | & | & | \\
C-C-C-C^*-C-C-C & & & C-C-C-C-C-C-C^*-C-C \\
\end{array}
$$

$$\text{(A) } D_{\delta\beta\beta'} \qquad\qquad\qquad \text{(B) } D_{\zeta\delta\beta} \tag{4.3}$$

Thus, for the (*mmmmm*) hexad, the values of $2D_{\delta\beta\beta'}(mm) + 2D_{\zeta\delta\beta}(mm)$ should be added.

$$
\begin{array}{ccccccc}
C^\zeta & C^\delta & C^\beta & C^\beta & C^\delta & C^\zeta \\
| & | & | & | & | & | \\
C-C-C-C-C-C-C^*-C-C-C-C-C-
\end{array}
\tag{4.4}
$$

The complete list of parameters is given in Table 4.4. The chemical shift, δ_{CH_2}, for the methylene carbon in question of the *mmmmm* hexad (C^* in Eq. 4.4) can be made in the following way:

$$\sum_i A_i = 2A_\beta + 2A_\delta + 2A_\zeta = 2 \times 7.57 + 2 \times 0.4 + 2 \times 0 = 15.94, \tag{4.5}$$

$$\sum_{ij} B_{ij} = 2B_{\beta\delta}(m) + 2B_{\delta\zeta}(m) = 2 \times (-0.34) + 2 \times 0 = -0.68, \tag{4.6}$$

$$\sum_{ij} C_{ij} = C_{\beta\beta}(m) + 2C_{\beta\delta}(m) + 2C_{\beta\zeta}(m) = 0.54 + 2 \times (-0.04) + 2 \times 0 = 0.46, \tag{4.7}$$

$$\sum_k D_k = 2D_{\delta\beta\beta'}(mm) + 2D_{\zeta\delta\beta}(mm) = 2 \times 0.516 + 2 \times (-0.164) = 0.704. \tag{4.8}$$

Then

$$\delta_{CH_2} = 29.9 + 15.94 - 0.68 + 0.46 + 0.704 = 46.32 \text{ ppm}. \tag{4.9}$$

Table 4.4. Empirical parameters for the estimation of ^{13}C shifts of the methylene group (Eq. 4.2) [9]. $D_{\delta\beta\beta'}$: $mm=0.516$, $mr=0.196$, $rm=0.018$, $rr=0.007$. $D_{\zeta\delta\beta}$: $mm=-0.164$, $mr=0.152$, $rm=-0.006$, $rr=0.004$

A_i	β	γ	δ	ε	ζ		
	7.57	−2.56	0.40	0.05	0		
B_{ij}	$B_{\beta\gamma}$	$B_{\beta\delta}$	$B_{\beta\varepsilon}$	$B_{\gamma\delta}$	$B_{\gamma\varepsilon}$	$B_{\delta\varepsilon}$	$B_{\delta\zeta}$
m	−1.5	−0.34	−0.25	0.4	−0.16	0	0
r	0.3	0.51	0	−0.1	−0.10	0	0
C_{ij}	$C_{\beta\beta}^a$	$C_{\beta\gamma}$	$C_{\beta\delta}$	$C_{\beta\varepsilon}$	$C_{\beta\zeta}$	$C_{\gamma\gamma}$	$C_{\gamma\delta}$
m	0.54	−0.001	−0.04	−0.107	0	−0.023	−0.017
r	0.34	−0.059	0	−0.072	0	0	0

[a] To be counted only once when used.

Table 4.5. Parameters for the estimation of ^{13}C shifts of the methyl group (Eq. 4.10) [9]

A_i	γ	δ	ε	ζ		
m	−3.05	0.77	0.16	0.05		
r	−4.83	0.22	0.06	−0.03		

B_{ij} & C_{ij}	$B_{\delta\zeta}$	$B_{\delta\varepsilon}$	$B_{\gamma\delta}$	$B_{\gamma\varepsilon}$	$B_{\varepsilon\zeta}$	$C_{\gamma\gamma}{}^{a}$
mm	0.02	0.17	0.75	−0.12	0	−1.7
mr	−0.10	−0.06	0.75	−0.40	0	0
rm	−0.15	−0.33	0.05	−0.14	0	0
rr	−0.20	−0.39	0.05	−0.52	0	1.1

^{a} To be counted only once when used.

The observed value measured in 1,2,4-trichlorobenzene at 120 °C is 46.52 ppm, which agrees well with the calculated value [10].

The shift equations for methyl and methine carbons are given as Eqs. (4.10) and (4.11) and the lists of parameters in Tables 4.5 and 4.6, respectively. Here, A_i, B_{ij}, and C_{ij} have the same meanings as in Eq. (4.2):

$$\delta_{CH_2} = 19.99 + \sum_i A_i + \sum_{ij} B_{ij} + \sum_{ij} C_{ij}, \tag{4.10}$$

$$\delta_{CH_2} = 33.26 + \sum_i A_i + \sum_{ij} B_{ij} + \sum_{ij} C_{ij}, \tag{4.11}$$

Equations (4.2), (4.10), and (4.11) should give the chemical shifts for carbons in the middle of a macromolecule, δ_{chain}. For carbons near the chain ends, the truncation effect should be taken into account and correction factors for this purpose are reported [9].

Table 4.6. Empirical parameters for the ^{13}C shifts of methines (Eq. 4.11) [9]

A_i	β	γ	δ	ε		
m	4.9	−2.38	0.36	0.04		
r	4.2	−2.46	0.27	0.00		

B_{ij} & C_{ij}	$B_{\gamma\varepsilon}$	$B_{\gamma\delta}$	$B_{\beta\gamma}$	$B_{\beta\delta}$	$B_{\delta\varepsilon}$	$B_{\beta\varepsilon}$	$C_{\beta\beta}{}^{a}$
mm	0.043	0.2	−0.8	−0.5	0	−0.01	0.8
mr	0.033	0.02	−1.4	−0.1	0	−0.01	−0.5
rm	−0.057	0	−1.0	−0.1	0	0	−0.5
rr	−0.078	0	−1.5	−0.4	0	0	−1.2

^{a} To be counted only once when used.

4.3 Monomer Reactivity Ratios and Statistical Treatments

The copolymerization reaction is usually described by monomer reactivity ratios r_1 $(=k_{11}/k_{12})$ and r_2 $(=k_{22}/k_{21})$. Here, k_{ij} is the rate constant for the following elementary reactions, for example, k_{12} is the rate constant for the reaction in Eq. (4.13).

$$\text{〜〜〜〜M}_1{}^* + \text{M}_1 \xrightarrow{k_{11}} \text{〜〜〜〜M}_1\text{M}_1{}^* \qquad (4.12)$$

$$\text{〜〜〜〜M}_1{}^* + \text{M}_2 \xrightarrow{k_{12}} \text{〜〜〜〜M}_1\text{M}_2{}^* \qquad (4.13)$$

$$\text{〜〜〜〜M}_2{}^* + \text{M}_1 \xrightarrow{k_{21}} \text{〜〜〜〜M}_2\text{M}_1{}^* \qquad (4.14)$$

$$\text{〜〜〜〜M}_2{}^* + \text{M}_2 \xrightarrow{k_{22}} \text{〜〜〜〜M}_2\text{M}_2{}^* \qquad (4.15)$$

r_1 and r_2 can describe the sequence statistics of copolymerization when the reactivity of the growing chain depends only on the nature of the monomer unit at the growing end; the chain statistics are first-order Markovian. When the reactivity of the growing chain is dependent on the natures of the ultimate and penultimate monomer units, four parameters of reactivity ratios r_{11} $(=k_{111}/k_{112})$, r_{21} $(=k_{211}/k_{212})$, r_{12} $(=k_{122}/k_{121})$, and r_{22} $(=k_{222}/k_{221})$ are necessary to describe the chain statistics, where, for example, k_{212} is the rate constant for the elementary reaction (Eq. 4.16), etc.

$$\text{〜〜〜M}_2\text{M}_1 + \text{M}_2 \xrightarrow{k_{212}} \text{〜〜〜M}_2\text{M}_1\text{M}_2{}^* \qquad (4.16)$$

In the previously mentioned round robin assessment of poly(MMA–co–AN)s with different compositions using 45 NMR spectrometers (Sect. 4.1), the r_1 and r_2 values were determined by the Fineman–Ross method for each result from the copolymer compositions analyzed by ^{1}H NMR spectroscopy [1]. The r_1 and r_2 values thus obtained were 1.38 and 0.32, respectively, on average. Rather large standard deviations were observed for the r_1 and r_2 as 23.6 and 21.8%, respectively.

The monomer reactivity ratios r_1 and r_2 in copolymerization can be determined from a single sample of copolymer as long as the mole fraction of each diad is obtained [11, 12]:

$$r_1 = \frac{[M_2]_0}{[M_1]_0} \frac{2(M_1M_1)}{(M_1M_2)}, \qquad (4.17)$$

$$r_2 = \frac{[M_1]_0}{[M_2]_0} \frac{2(M_2M_2)}{(M_1M_2)}, \qquad (4.18)$$

^{13}C NMR spectra of the copolymer of MMA and AN are sensitive to the monomer sequence distribution as indicated in Fig. 4.6. The peak assignments for the MMA-centered and AN-centered triads were made according to the literature [13]. Further splittings may be ascribed to the configurational sequences. The fractions

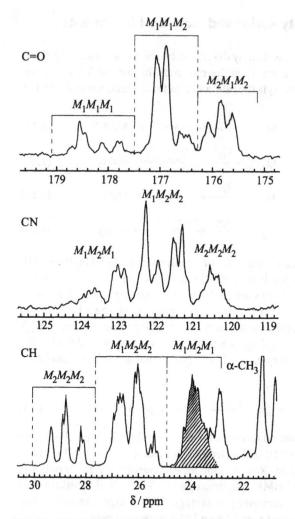

Fig. 4.6. 100 MHz ^{13}C NMR spectra of poly(MMA–*co*–AN) (sample 2 in Table 4.7). All the signals are sensitive to triads of the monomer sequence as indicated in the figure (M_1=MMA, M_2=AN) in acetonitrile-d_3 at 70 °C [1]. In the CH region spectrum the *hatched area* is assigned to the $M_1M_2M_1$ triad according to the literature [13]

of the six triads of monomer sequences were determined from the carbonyl (MMA=M_1) and methine (AN=M_2) carbon signals as shown in Table 4.7.

The diad fractions can be derived from the triad fractions thus determined, as shown in the following equations:

$$(M_1M_1) = (M_1M_1M_1) + (M_1M_1M_2)/2, \tag{4.19}$$

$$(M_1M_2) = (M_2M_1M_2) + (M_1M_2M_1) + (M_1M_1M_2)/2 + (M_1M_2M_2)/2, \tag{4.20}$$

$$(M_2M_2) = (M_2M_2M_2) + (M_1M_2M_2)/2 \tag{4.21}$$

Table 4.7. Analysis of monomer sequence distribution in poly(MMA-*co*-AN) by ^{13}C NMR spectroscopy. The figures in *parentheses* represent the standard deviation σ (%) [1]

Number	MMA content		MMA (M_1)-centered triad (%)				AN (M_2)-centered triad (%)			
	Feed	copolymer[a]	n^b	$M_1M_1M_1$	$M_1M_1M_2$	$M_2M_1M_2$	n^b	$M_1M_2M_1$	$M_1M_2M_2$	$M_2M_2M_2$
1	0.206	0.375	45	2.9 (29.2)	15.6 (4.2)	19.0 (5.0)	46	11.9 (5.5)	30.6 (1.8)	20.0 (3.2)
2	0.347	0.520	45	8.9 (6.8)	26.8 (2.5)	16.3 (7.0)	46	17.8 (4.0)	23.0 (2.4)	7.2 (6.9)
3	0.548	0.671	44	23.5 (6.0)	34.2 (3.4)	9.4 (11.3)	45	19.2 (4.5)	11.8 (5.3)	1.9 (22.3)

[a] Average data of ^{1}H NMR analysis.
[b] Number of determinations.

Table 4.8. Reactivity ratios r_i and r_{ij} for the copolymerization of MMA (M_1) and AN (M_2) in DMSO at 50 °C. The figures in *parentheses* represent the standard deviation σ (%) [1]

Number[a]	r_1	r_2	r_{11}	r_{21}	r_{12}	r_{22}
1	1.55	0.34	1.44 (26.8)	1.56 (7.9)	0.34 (7.0)	0.34 (4.8)
2	1.41	0.34	1.24 (13.9)	1.55 (8.1)	0.34 (5.7)	0.33 (7.2)
3	1.27	0.38	1.13 (9.4)	1.52 (11.6)	0.38 (10.1)	0.39 (23.1)

[a] The number of the sample is the same as that in Table 4.7.

Thus the r_1 and r_2 values were calculated using Eqs. (4.17) and (4.18) and are shown in Table 4.8. The monomer reactivity ratios, particularly r_1, clearly depend on the copolymer compositions. This indicates that the copolymerization of MMA and AN cannot be described with the terminal model, as reported in the literature [12–14]. This may be one of the reasons for the large standard deviations for the previously mentioned r_1 and r_2 values determined from Fineman–Ross plots.

The penultimate-model reactivity ratios, r_{ij}, can be determined from triad monomer sequence distribution data using Eqs. (4.22)–(4.25):

$$r_{11} = \frac{[M_2]_0}{[M_1]_0} \frac{2(M_1M_1M_1)}{(M_1M_1M_2)},$$
(4.22)

$$r_{21} = \frac{[M_2]_0}{[M_1]_0} \frac{(M_1M_1M_2)}{2(M_2M_1M_2)},$$
(4.23)

$$r_{12} = \frac{[M_1]_0}{[M_2]_0} \frac{(M_1M_2M_2)}{2(M_1M_2M_1)},$$
(4.24)

$$r_{22} = \frac{[M_1]_0}{[M_2]_0} \frac{2(M_2M_2M_2)}{(M_1M_2M_2)},$$
(4.25)

where M_1 is MMA and M_2 is AN [11, 12].

Averaged r_{ij} values for the three copolymer samples are shown in Table 4.8. Most of the σ values for r_{ij} were around 10%. Larger σ values for r_{11} of sample 1 and for r_{22} of sample 3 could be ascribed to smaller peak intensities for the $M_1M_1M_1$ triad and the $M_2M_2M_2$ triad, respectively. The r_{11} value is different from the r_{21} value for all samples. This is a clear indication of a significant penultimate-group effect characteristic of the chain-end MMA radical. Both the r_{11} and r_{21} values, particularly the former, decreased with an increase in the content of MMA in the copolymer. This suggests the possible existence of a prepenultimate effect in the reactivity of $\sim\sim\sim M_1\cdot$ radicals.

On the other hand, the values of r_{12} and r_{22} determined for each copolymer were close to each other, both being enhanced slightly with increasing MMA content

Table 4.9. Penultimate model reactivity ratios r_{ij} in the copolymerization of MMA (M_1) and AN (M_2) in DMSO at 50 °C determined from comonomer sequence analysis. Copolymerizations were carried out in DMSO at 50 °C and at $[MMA+AN]_0/[AIBN]=150$ except for samples 10–12. For samples 10–12, see footnotes a–c [1]

Sample	MMA content		r_{11}	r_{21}	r_{12}	r_{22}
	Feed	Copolymer				
4	0.205	0.365	1.49	1.58	0.32	0.32
5	0.347	0.515	1.32	1.46	0.36	0.33
6	0.398	0.564	1.23	1.43	0.34	0.34
7	0.453	0.610	1.13	1.49	0.35	0.36
8	0.508	0.645	1.09	1.36	0.37	0.38
9	0.548	0.679	1.12	1.45	0.37	0.45
10[a]	0.548	0.678	1.08	1.39	0.36	0.41
11[b]	0.548	0.675	1.06	1.46	0.38	0.43
12[c]	0.548	0.672	1.07	1.41	0.39	0.42

[a] $[MMA+AN]_0/[AIBN]_0 = 75$.
[b] Polymerization temperature 60 °C.
[c] Polymerization temperature 40 °C.

in the copolymer. Such a compositional dependence of r_{12} and r_{22} was also found in additional experiments on another series of copolymers having a wider range of composition (Table 4.9). The increase in r_{12} and r_{22} was very clearly evidenced especially in the compositional range with higher MMA contents. The results indicate that there is a prepenultimate effect but little penultimate effect in the reactivity of the chain-end AN radical in this copolymerization (see r_2 values in Table 4.8). The prepenultimate effect with little penultimate one may be caused by a peculiar structure of the propagating radicals, such as the radicals stabilized by the prepenultimate substituent.

The fluctuation of monomer reactivity ratios is affected by the fluctuation of the results of the copolymerization. In order to study the reproducibility and precision of the results of the copolymerization, additional runs of the copolymerization were made in Hatada's laboratory under various conditions, including those adopted for the round robin copolymer samples 1–3 and ^{13}C NMR analyses of these copolymers were performed with a 125 MHz NMR spectrometer. The penultimate-model reactivity ratios obtained from the ^{13}C NMR spectra of these copolymer samples are shown in Table 4.9. The analyses of three copolymer samples 4, 5, and 9 in the table, each of which was prepared under the same conditions as for samples 1, 2, and 3, respectively, showed the reactivity ratios very close to those of the corresponding samples among the latter series of copolymers (Table 4.8). This is clear indication of the excellent reproducibility of the polymerization reaction as well as of the evaluation of the monomer reactivity ratios. All the reactivity ratios for six different copolymers (samples 4–9 in Table 4.9) depend on the copolymer compositions; the r_{11} and r_{21} values decrease and the r_{12} and r_{22} values increase slightly with an increase in the MMA content of the copolymer. The results are consistent with those shown in Table 4.8.

As shown in Table 4.9, the temperature of copolymerization scarcely affects the results of copolymerization at least in the temperature range of 40–60 °C (samples 9, 11, 12). The ratio of monomer and initiator concentrations (samples 9, 10) also does not affect the results. This clearly indicates that the fluctuation of the temperature or the concentration of the reagents, even if it exists, hardly affects the values of the monomer reactivity ratios. Then, all the results in Table 4.9 show the reliability of the results in Table 4.8. However, the precision and accuracy of the analytical results for polymer and copolymer inevitably include those of the polymerization reactions. This should always be kept in mind in the course of polymer characterization.

4.4 Analysis of Diene Polymers

Polymerization of diene monomers, such as 1,3-butadiene and isoprene, can produce polymers with different structures of monomeric units, 1,4 and 1,2 enchainments; the former can produce *cis* and *trans* configurational isomeric structures, and the latter leads to structures with the same configurational properties as vinyl polymers, i.e. tacticity (Chap. 3). Thus the structural analysis of diene polymers should involve structural analyses of composition, configuration, and sequence distribution as in the case of copolymers.

The 100 MHz ^{1}H NMR spectrum of polybutadiene prepared with n-C_4H_9Li/ $O(C_2H_5)_2$ in a hydrocarbon composed of 1,4 and 1,2 structures (53% 1,2 structure) is shown in Fig. 4.7 [15]. The 1,4 and 1,2 structures can be distinguished by observing olefinic proton signals; vinylidene methylene proton signals at 4.9 ppm (signal C) due to the side group in the 1,2 unit, and vinylene methine proton signals at 5.4 ppm (signal B) due to the main-chain methine group in 1,4-units. The assignment of signal C was made by taking account of the spectrum of poly-(butadiene-1,1,4,4,-d_4) in which no signals appeared around 4.9 ppm. The assignment for signal B was made by comparing the spectrum with those of *cis*-1,4-polybutadienes and *trans*-1,4-polybutadienes, the structures of which were evidenced by X-ray analysis for the solid polymers. On the basis of the relative intensities (I_B and I_C) of the two olefinic proton signals around 4.9–5.4 ppm, the contents of the 1,2 and 1,4 units can be calculated by the following equations:

$$1,2\text{-structure}(\%) = I_C/(I_B + I_C) \times 100, \tag{4.26}$$

$$1,4\text{-structure}(\%) = I_B/(I_B + I_C) \times 100. \tag{4.27}$$

Analyses of configuration and sequence distribution of monomeric units can be made by using ^{13}C NMR spectra. The 50 MHz ^{13}C NMR spectra of saturated carbons

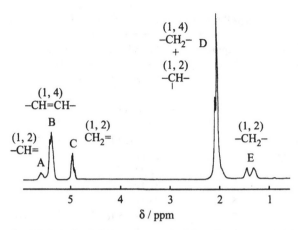

Fig. 4.7. 100 MHz ^{1}H NMR spectrum of polybutadiene prepared with n-C$_4$H$_9$Li/O(C$_2$H$_5$)$_2$ in hydrocarbon measured in CDCl$_3$ at 60 °C [15]

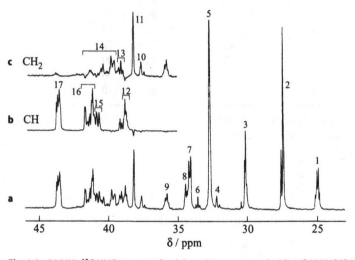

Fig. 4.8. 50 MHz ^{13}C NMR spectra of polybutadiene prepared with n-C$_4$H$_9$Li/O(C$_2$H$_5$)$_2$ in hydrocarbon measured in CDCl$_3$ at 35 °C: **a** normal spectrum; **b, c** spectra obtained by the DEPT method [16]

of the polybutadiene prepared with n-C$_4$H$_9$Li/O(C$_2$H$_5$)$_2$ are shown in Fig. 4.8 [16]. In the chemical shift range of 37–44 ppm, methine and methylene carbon signals overlap with each other and the signals of methine and methylene carbons were separately measured by the DEPT method.[3] The assignments of the signals

---

[3] By using the distortionless enhancement of polarization transfer (DEPT) technique with appropriate parameters in the pulse sequence, one can differentiate CH$_3$, CH$_2$, and CH carbon signals [17]. The spectra of CH and CH$_2$ carbons obtained by the DEPT method are shown in Fig. 4.8b and c, respectively.

Table 4.10. Peak assignments for the[13]C NMR spectrum of polybutadiene [16]

Sequence[a]	Carbon[b]	Signal[c]	Chemical shift (ppm)
C–v	4	1	24.98–25.10
C–(1,4)	4	2	27.42–27.57
(1,4)–C	1	2	27.42–27.57
T–v	4	3	30.16
v–v–C (m)	1	4	31.60–32.13
(1,4)–v–C	1	5	32.72
T–(1,4)	4	5	32.72
(1,4)–T	1	5	32.72
v–v–C (r)	1	6	33.35–33.53
(1,4)–V–(1,4)	1	7	33.99–34.16
(1,4)–V–v (m)	1	8	34.31
(1,4)–V–v (r)	1	9	35.63–36.00
v–v–T (m)	1	10	37.24–37.48
(1,4)–v–T	1	11	38.18
v–V–v	2	12	38.57–39.13
v–v–T (r)	1	13	38.96–39.13
v–V	1	14	39.43–41.72
(1,4)–V–v	2	15	40.55–41.00
v–V–(1,4)	2	16	41.00–41.66
(1,4)–V–(1,4)	2	17	43.47–43.70

[a] C: cis-1,4 unit; T: trans-1,4 unit; V: 1,2 unit; (1,4): C+T, m: meso; r: racemo. For example, C–v represents the cis-1,4 unit in structure.

and v–v–C (m) represents the *cis*-1,4 unit in the structure.

[b] Number for carbons in the structural unit represented by the capital letters C, T, and V.

[c] See Fig. 4.8.

were made by using model compounds that correspond to mono-ad and diad sequences of each isomeric structure, considering the fractions of isomeric structures calculated from the results of the ^{1}H NMR spectrum. The assignments thus obtained are shown in Table 4.10. The fine splittings of each peak are due to the tacticity of the unit and/or the isomeric structures of the neighboring units.

The diad sequence distribution of cis-1,4 (C), trans-1,4 (T), and 1,2 units (V) can be calculated from methylene carbon signals by using following equations, where, for example, [C-V] represents the fraction of (cis-1,4)–(1,2). I_i represents the intensity of the signal i and TI (total intensity) is the calculated value from Eq. (4.35) (Table 4.10).

$$[C\text{-}V] = I_1/\text{TI}, \tag{4.28}$$

$$[T\text{-}V] = I_3/\text{TI}, \tag{4.29}$$

$$[V\text{-}C] = [C\text{-}V] + [T\text{-}V] - [V\text{-}T], \tag{4.30}$$

$$[V\text{-}T] = (I_{10} + I_{11} + I_{13})/\text{TI}, \tag{4.31}$$

$$[C\text{-}(1,4)] = (1/2)I_2/\text{TI}, \tag{4.32}$$

$$[T\text{-}(1,4)] = (1/2)\{(I_4 + I_5 + I_6)/\text{TI} - [V - C]\}, \tag{4.33}$$

$$[V\text{-}V] = I_{14}/\text{TI}, \tag{4.34}$$

$$\text{TI} = (3/2)\,(I_1 + I_3) + (1/2)\,(I_2 + I_4 + I_5 + I_6 + I_{10} + I_{11} + I_{13}) + I_{14}. \tag{4.35}$$

The values of the NOE (2.7–3.0) are almost equal for all the signals, and spin–lattice relaxation times are less than 1 s for the measurements at 50 MHz at 35 °C. So accurate measurements of the signal intensities are easy to make.

The 100 MHz ^{1}H NMR spectrum of polyisoprene prepared with n-C$_4$H$_9$Li/ O(C$_2$H$_5$)$_2$ in hexane is shown in Fig. 4.9 [18]. The signals could be assigned as indicated in the figure by using polymers of isoprene-1,1,4,4-d_4, isoprene-1,1,5,5,5-d_5, and isoprene-4,4-d_2. The peak assignments for the cis-1,4 and trans-1,4 structures (peaks F and G) were made by using cis-1,4-poly(isoprene-1,1,5,5,5-d_5) and trans-1,4-poly(isoprene-1,1,5,5,5-d_5) prepared with TiCl$_4$/Al(C$_2$H$_5$)$_3$ and VCl$_3$/Al(C$_2$H$_5$)$_3$, respectively. The isomeric structures of the polymers were analyzed by X-ray spectroscopy. On the basis of the relative intensities of the olefinic proton signals around 5 ppm, the fractions of the 1,4-, 3,4-, and 1,2- isomeric structures can be calculated using Eqs. (4.36)–(4.38). The fractions of the cis-1,4 and trans-1,4 structures can be obtained by Eqs. (4.39) and (4.40).

$$1,4\ \text{structure(\%)} = 100 \times I_B/(I_B + I_C/2 + I_D/2), \tag{4.36}$$

$$3,4\ \text{structure(\%)} = 100 \times (I_D/2)/(I_B + I_C/2 + I_D/2), \tag{4.37}$$

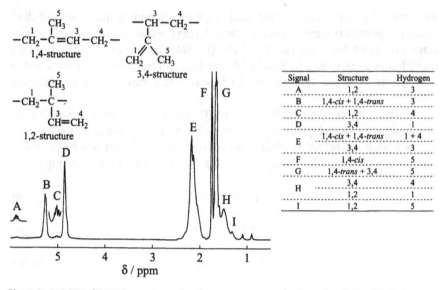

Signal	Structure	Hydrogen
A	1,2	3
B	1,4-*cis* + 1,4-*trans*	3
C	1,2	4
D	3,4	1
E	1,4-*cis* + 1,4-*trans*	1 + 4
	3,4	3
F	1,4-*cis*	5
G	1,4-*trans* + 3,4	5
H	3,4	4
	1,2	1
I	1,2	5

Fig. 4.9. 100 MHz ^{1}H NMR spectrum of polyisoprene prepared with n-C$_4$H$_9$Li/O(C$_2$H$_5$)$_2$ in hexane measured in CDCl$_3$ at 60 °C [18]

$$1,2 \text{ structure(\%)} = 100 \times (I_C/2)/(I_B + I_C/2 + I_D/2), \tag{4.38}$$

$$cis\text{-}1,4 \text{ structure(\%)} = [I_F/(I_F + I_G)] \times [1,4 \text{ structure(\%)} + 3,4 \text{ structure(\%)}], \tag{4.39}$$

$$trans\text{-}1,4 \text{ structure(\%)} = 1,4 \text{ structure(\%)} - cis\text{-}1,4 \text{ structure(\%)}. \tag{4.40}$$

Natural rubber is almost 100% *cis*-1,4-polyisoprene and shows a very simple ^{13}C NMR spectrum. On the other hand, polyisoprenes that contain *cis*-1,4, *trans*-1,4, and 3,4 units exhibit complicated ^{13}C NMR spectra, resulting from various types of carbon atoms in different sequences. The polyisoprene prepared with a radical initiator contains head-to-head and tail-to-tail structures in addition to regular sequences of 1,4 and 3,4 structures. To accomplish the peak assignments for the ^{13}C NMR spectra of these polymers is very difficult; hydrogenation of the polymers [19] is an efficient means and provides information about the sequence distribution in the polyisoprenes. Through hydrogenation *cis*-1,4 and *trans*-1,4 units are converted into the same structure (I), and the 3,4 unit is converted into structure II. Thus, hydrogenated polyisoprenes that contain almost no 1,2 units can be regarded as copolymers consisting of structures I and II.

$$\underset{\substack{cis\text{-}1,4 \text{ and } trans\text{-}1,4}}{-CH_2-\overset{\overset{\displaystyle CH_3}{|}}{C}=CH-CH_2-} \xrightarrow{\text{hydrogenation}} \underset{\text{structure (I)}}{-C^1H_2-\overset{\overset{\displaystyle C^5H_3}{|}}{C^2}H-C^3H_2-C^4H_2-}$$

$$(4.41)$$

$$-CH-CH_2- \xrightarrow{\text{hydrogenation}} -C^3H-C^4H_2-$$

3,4-unit   structure (II)   (4.42)

The chemical shift of each carbon atom in the hydrogenated polymer can be predicted by using the empirical equations for branched and linear alkanes proposed by Paul and Grant [6] or Lindeman and Adams [7].

The 25 MHz ^{13}C NMR spectrum of hydrogenated polyisoprene prepared with n-$C_4H_9Li/O(C_2H_5)_2$ is shown in Fig. 4.10 [20, 21]. The polymer exhibited 16 main signals as indicated by symbols A–P. The chemical shifts of the carbon atoms in the central monomer units of triad sequences of structures I and II were calculated by using the equation of Lindeman and Adams [7], and were compared with observed ones to assign the signals. The fractions of the 3,4 units in the starting polymers are also taken into account for the assignments. The results are listed in Table 4.11. The fractions of the 1,4-centered triads were determined from signals H, G, and F according to the following equations. The calculated chemical shifts for the C^2 carbon in 1,4–1,4–3,4 and 3,4–1,4–1,4 triads are equal to each other and the peak separation that corresponds to the two kinds of carbons could not be detected. So the two triads cannot be distinguished from each other in the spectrum in Fig. 4.10.

$$(1,4)-(1,4)-(1,4)\text{fraction} = I_H/(I_H + I_G + I_F) \times 1,4 \text{ unit}, \tag{4.43}$$

$$(1,4)-(1,4)-(3,4)\text{fraction} + (3,4)-(1,4)-(1,4)\text{fraction}$$
$$= I_G/(I_H + I_G + I_F) \times 1,4 \text{ unit}, \tag{4.44}$$

$$(3,4)-(1,4)-(3,4)\text{fraction} = I_F/(I_H + I_G + I_F) \times 1,4 \text{ unit}. \tag{4.45}$$

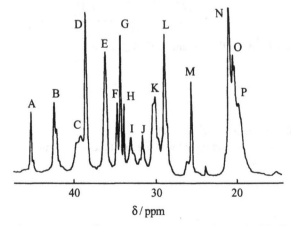

Fig. 4.10. 25 MHz ^{13}C NMR spectrum measured in $CDCl_3$ of the hydrogenated polyisoprene prepared with n-$C_4H_9Li/O(C_2H_5)_2$ [20, 21]

Table 4.11. Peak assignments of hydrogenated polyisoprene for the ^{13}C NMR spectrum [21]

Triad sequence	Carbon[a]	Signal[b]	Chemical shift (ppm)	
			Observed[b]	Calculated[c]
(1,4)–(<u>1,4</u>)–(1,4)	1	D	37.47	37.16
	2	H	32.84	32.52
	3	D	37.47	37.16
	4	M	24.51	24.58
	5	N	19.83	19.80
(1,4)–(<u>1,4</u>)–(3,4)	1	D	37.47	37.16
	2	G	33.25	32.52
	3	E	35.09	34.22
	4	L	27.84	28.84[d], 29.09
	5	N	19.83	19.80
(3,4)–(<u>1,4</u>)–(1,4)	1	E	35.09	34.22
	2	G	33.25	32.52
	3	D	37.47	37.16
	4	M	24.51	24.58
	5	N	19.83	19.80
(3,4)–(<u>1,4</u>)–(3,4)	1	E	35.09	34.22
	2	F	33.64	32.52
	3	E	35.09	34.22
	4	L	27.84	28.84[d], 29.09
	5	N	19.83	19.80
(1,4)–(<u>3,4</u>)–(1,4)	1, 5	O	19.34	19.80
	2	K	29.32	30.46
	3	A	44.39	43.66
	4	L	27.84	28.84

[a] Each carbon atom is denoted as follows: for the 1,4 unit $-C^1-\overset{\overset{\displaystyle C^5}{|}}{C^2}-C^3-C^4-$,

for the 3,4 unit. $\overset{\displaystyle -C^3-C^4-}{\underset{\overset{\displaystyle C^2}{\diagup\,\diagdown}}{}}$ $C^1\quad C^5$

[b] See Fig. 4.10.

[c] Calculated using the equation of Lindeman and Adams [7].

[d] Calculated chemical shift varies with the structure of the subsequent monomer unit.

Table 4.11 (continued)

Triad sequence	Carbon[a]	Signal[b]	Chemical shift (ppm)	
			Observed[b]	Calculated[c]
(1,4)-(3,4)-(3,4)	1, 5	P	18.61	19.80
	2	K	29.32	30.46
	3	B	41.43	41.59
	4	I	31.96	33.35[d], 33.60
	4	J	30.58	33.35[d], 33.60
(3,4)-(3,4)-(1,4)	1, 5	P	18.61	19.80
	2	K	29.32	30.46
	3	B	41.43	41.59
	4	L	27.84	28.84
(3,4)-(3,4)-(3,4)	1, 5	P	18.61	19.80
	2	K	29.32	30.46
	3	C	38.13	39.52
	4	I	31.96	33.60[d], 33.85
	4	J	30.58	33.60[d], 33.85

The fractions of the 3,4-centered triad sequences can be determined from signals A, B, and C in Fig. 4.10 that are due to carbon C^3 (see Eq. 4.42, Table 4.11). However, signal C overlaps with signal D and it is difficult to measure directly the intensity of signal C. The intensity of signal C was determined from the intensities of signals A, B, and K by using Eq. (4.46):

$$I_A + I_B + I_C = I_K . \tag{4.46}$$

Hence the fractions of the triad sequences of the 3,4 unit were determined from Eqs. (4.47)–(4.49):

$$(1,4)-(3,4)-(1,4)\text{fraction} = I_A/I_K \times 3,4 \text{ unit}, \tag{4.47}$$

$$(1,4)-(3,4)-(3,4)\text{fraction} + (3,4)-(3,4)-(1,4)\text{fraction}$$
$$= I_B/I_K \times 3,4 \text{ unit}, \tag{4.48}$$

$$(3,4)-(3,4)-(3,4)\text{fraction} = [1-(I_A + I_B)/I_K] \times 3,4 \text{ unit}. \tag{4.49}$$

Again the 1,4–3,4–3,4 and 3,4–3,4–1,4 triads cannot be distinguished from each other.

The amount of the 3,4 unit could also be determined from the signals due to the tertiary carbon C^2. In Fig. 4.10, the resonance of C^2 in structure I splits into three peaks, signals F, G, and H, depending on the structure of adjacent units, and the

Table 4.12. Microstructure of polyisoprenes prepared with n-$C_4H_9Li/O(C_2H_5)_2$ as determined by ^1H and ^{13}C NMR spectroscopies (polymerization conditions: isoprene 2.5 mol/l, n-C_4H_9Li 1.25×10^{-2} mol/l) [20, 21]

Polymer	$O(C_2H_5)_2/n$-C_4H_9Li	Fraction of 3,4 unit	
		^1H NMR	^{13}C NMR
1	0.0	0.045	0.052
2	1.0	0.170	0.188
3	2.0	0.218	0.219
4	4.7	0.339	0.349
5	11.8	0.430	0.430
6	23.7	0.450	0.480

resonance of C^2 in structure II appears as signal K regardless of the structure of the neighboring units. The amount of structure II, which corresponds to the content of the 3,4 unit in the starting polymer, was obtained by using the following equation:

$$3,4 \text{ unit} = I_K/(I_F + I_G + I_H + I_K). \tag{4.50}$$

The amounts of the 3,4 unit thus obtained were in good agreement with those obtained by ^1H NMR analysis as shown in Table 4.12. This clearly indicates the validity of the assignments in the ^{13}C NMR spectra. It should be noted that the NOE values are almost the maximum independent of the type of carbon in the 25 MHz ^{13}C NMR spectra (Sect. 7.4).

References

1. HATADA K, KITAYAMA T, TERAWAKI Y, SATO H, CHÛJÔ R, TANAKA Y, KITAMARU R, ANDO I, HIKICHI K, HORII F, members of research group on NMR SPSJ (1995) Polym J 27:1104
2. BOVEY FA (1972) High resolution NMR of macromolecules. Academic, New York, pp 205, 219
3. HELLWEGE KH, JOHNSON U, KOLBE K (1966) Kolloid Z Z Polym 214:45
4. NISHIURA T, KITAYAMA T, HATADA K (2000) Int J Polym Anal Charact 5:401
5. KITAYAMA T, UTE K, YAMAMOTO M, FUJIMOTO N, HATADA K (1991) Polym Bull 25:683
6. PAUL EG, GRANT DM (1964) J Am Chem Soc 86:2984
7. LINDEMAN LP, ADAMS JQ (1971) Anal Chem 43:1245
8. CARMAN CJ, TARPLEY AR JR, GOLDSTEIN JH (1973) Macromolecules 6:719
9. CHENG HN, BENNETT MA (1987) Makromol Chem 188:135
10. RANDALL JC (1977) Polymer sequence determination – carbon-13 NMR method. Academic, New York, p 15
11. CHÛJÔ R (1966) J Phys Soc Jpn 21:2669

12. Chûjô R, Ubara H, Nishioka A (1972) Polym J 3:670
13. Gerken T, Ritchey WM (1978) J Appl Polym Sci Appl Polym Symp 34:17
14. Johnsen VU, Kolbe K (1968) Macromol Chem 116:173
15. Tanaka Y, Takeuchi Y, Kobayashi M, Tadokoro H (1971) J Polym Sci A-2 9:43
16. Sato H, Takebayashi K, Tanaka Y (1987) Macromolecules 20:2418
17. Chandra Kumar N, Subramanian S (1987) Modern techniques in high-resolution FT-NMR. Springer, Berlin Heidelberg New York, p 86
18. Sato H, Tanaka Y (1979) J Polym Sci Polym Chem Ed 17:3551
19. Sanui K, Macknight WJ, Lentz RW (1973) J Polym Sci Polym Lett Ed 11:427
20. Tanaka Y, Sato H (1976) Polymer 17:413
21. Tanaka Y, Sato H, Ogura A (1976) J Polym Sci Polym Chem Ed 14:73

5 NMR for the Study of Polymerization Reactions

5.1 Analysis of End Groups and Structural Defects in Polymer Chains

In addition to the structural determination for the polymer main chain, such as tacticity and comonomer sequence as discussed in previous chapters (Chaps. 3, 4), analysis of end groups or abnormal linkages in polymer chains, which are very small in their weight fractions within the polymers, is a powerful technique for studies of the mechanism of polymerization [1, 2]. In the study of radical polymerization of vinyl monomers, ^{14}C-labeled initiators were used to investigate the number of initiator fragments attached to the polymer chain ends [3, 4]. The number reflects the extent of two possible termination reactions, combination and disproportionation, as far as chain-transfer reactions are negligible. When a polymer molecule is formed through combination, the chain contains two initiator fragments on both ends, while that formed through disproportionation contains one fragment (Structure 5.1).

Structure 5.1

Though the radioactive tracer method is very sensitive for the detection of the fragments derived from the initiator, the use of a ^{14}C-labeled initiator only gives information on the number of initiator fragments and is not effective for the study of how the fragment is incorporated into the polymer chain. Since FT NMR spectroscopy became the standard of NMR measurement, which allows us to accumulate FID signals to observe small signals in a sample, stable and NMR-sensitive isotope-labeled initiators have been used to study the initiation and termination steps, especially in radical polymerization and copolymerization. The NMR method enables us to determine not only the number per chain but also the structures of initiator fragments, which give us valuable information for the understanding of the mechanism of polymerization.

For example, when vinyl chloride is polymerized with AIBN containing ^{13}CN groups, the polymer formed exhibits CN signals due not only to the group of initiating chain end (Structure 5.2) but also to the in-chain unit (methacrylonitrile unit); the latter is incorporated by the copolymerization of methacrylonitrile formed through disproportionation (Structure 5.3) of the primary radicals [5]. The ^{14}C labeling method cannot distinguish these two types of initiator residues, and the number of AIBN fragments at the chain end should be overestimated, leading to misunderstanding of the polymerization mechanism.

Structure 5.2

Structure 5.3

The 750 MHz ^1H NMR spectrum of PMMA prepared with benzoyl peroxide (BPO) in toluene at 100 °C is shown in Fig. 5.1 [6]. The expanded spectra inserted in the figure show signals from end groups and those due to the terminal monomeric units. In the aromatic-region spectrum, signals at 7.95, 7.52, and 7.41 ppm were assigned to the benzoyloxy group (Fig. 5.1, range A) and those at 6.95–7.25 ppm to the phenyl group (Fig. 5.1, range B); splittings of the former as ortho, meta, and para protons and their chemical shifts are very typical for the benzoate group. Methylene protons of the first monomer unit adjacent to the benzoyloxy fragment show a signal at 4.10–4.45 ppm (Fig. 5.1, range D). The two protons of the methylene group are nonequivalent owing to the existence of the adjacent asymmetric carbon. They also show the splitting due to the tacticity of the first diad of the polymer chain. The signals consist of two sets of AB quartets with different intensities. The intensity ratio (m/r) of 41/59 differs from that in the chain, 25/75. This indicates that the stereospecificity of the initiation process in the radical polymerization differs from that in the propagating process. If all the benzoyloxy group is attached to the chain end as depicted by the structure I in Eq. (5.4), the intensity ratio of the benzoyloxy proton signals (Fig. 5.1, range A) and the methylene proton signals (Fig. 5.1, range D) is expected to be 5/2; however, the observed value was 5.82/2, suggesting the existence of another type of benzoyloxy group. The most probable structure is that of IV in Structure 5.7, which is derived from the primary radical

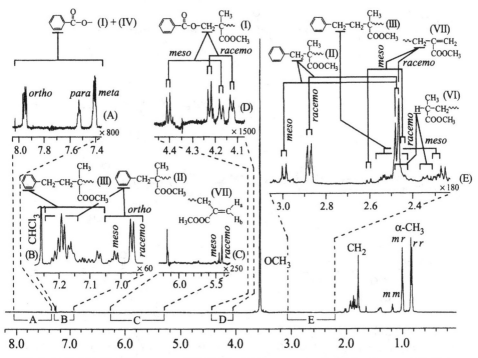

Fig. 5.1. 750 MHz ^{1}H NMR spectrum of PMMA ($\bar{M}_n$=17,700) prepared with BPO in toluene at 100 °C [6]. 10 wt/vol%, CDCl$_3$ solution, 55 °C, 45° pulse, pulse repetition time 18 s, 512 scans

termination occurring between the propagating radical and benzoyloxy radical generated from BPO (Structure 5.7) or from chain-transfer reaction of the propagating radical to BPO (Structure 5.8), or head addition of benzoyloxy radical to MMA in the initiation process (Structures 5.4–5.11).

Another aromatic region (Fig. 5.1, range B) is typical of alkylbenzenes and is assignable to the phenyl group attached to the chain end as in structure II in Structure 5.5 derived from the initiation by the phenyl radical which is generated from the benzoyloxy radical through decarboxylation. To confirm this, two PMMA samples having the phenyl group at the initiating chain end were prepared by anionic polymerization with phenylmagnesium bromide in THF and in toluene; the former is syndiotactic and the latter isotactic. The phenyl region of these PMMAs are shown in Fig. 5.2b and c along with the corresponding signals of the radically prepared PMMA (Fig. 5.2a). The peaks at 6.95–7.05 ppm are assigned to ortho protons as they split into doublets, and the peaks at 7.15–7.24 ppm are due to meta and para protons. The ortho proton signals are sensitive to the stereochemistry of the first diad and the split signals are assigned on the basis of the tacticity of the two previously mentioned syndiotactic and isotactic PMMAs as shown in the figure. The chemical shifts of the ortho protons for the meso diad in the syndiotactic PMMA (Fig. 5.2b) and in the isotactic one (Fig. 5.2c) are slightly different from each other, reflecting triad tacticity instead of diad as indicated in

$$\frac{1}{2}\ \mathrm{C_6H_5\text{-}\overset{O}{\overset{\|}{C}}\text{-}O\text{-}O\text{-}\overset{O}{\overset{\|}{C}}\text{-}C_6H_5} \longrightarrow C_6H_5\text{-}\overset{O}{\overset{\|}{C}}\text{-}O\bullet \xrightarrow{\ MMA\ } C_6H_5\text{-}\overset{O}{\overset{\|}{C}}\text{-}O\text{-}CH_2\overset{CH_3}{\underset{C=O,\ OCH_3}{\overset{|}{\underset{|}{C}}}}\sim\!\sim \qquad (I) \qquad 5.4$$

$$\downarrow\ -CO_2$$

$$C_6H_5\bullet \xrightarrow{\ MMA\ } C_6H_5\text{-}H_2C\text{-}\overset{CH_3}{\underset{C=O,\ OCH_3}{\overset{|}{\underset{|}{C}}}}\sim\!\sim \qquad (II) \qquad 5.5$$

chain transfer to toluene $\longrightarrow$ $C_6H_5\text{-}CH_2\bullet \xrightarrow{\ MMA\ } C_6H_5\text{-}CH_2\text{-}CH_2\text{-}\overset{CH_3}{\underset{C=O,\ OCH_3}{\overset{|}{\underset{|}{C}}}}\sim\!\sim$ (III)   5.6

$$\sim\!\sim CH_2\text{-}\overset{CH_3}{\underset{\underset{OCH_3}{C=O}}{\overset{|}{\underset{|}{C}}}}\bullet\ +\ \bullet O\text{-}\overset{O}{\overset{\|}{C}}\text{-}C_6H_5 \longrightarrow \sim\!\sim CH_2\text{-}\overset{CH_3}{\underset{\underset{H_3CO}{C=O}}{\overset{|}{\underset{|}{C}}}}\text{-}O\text{-}\overset{O}{\overset{\|}{C}}\text{-}C_6H_5 \quad (IV) \qquad 5.7$$

$$\sim\!\sim CH_2\text{-}\overset{CH_3}{\underset{\underset{OCH_3}{C=O}}{\overset{|}{\underset{|}{C}}}}\bullet\ +\ C_6H_5\text{-}\overset{O}{\overset{\|}{C}}\text{-}O\text{-}O\text{-}\overset{O}{\overset{\|}{C}}\text{-}C_6H_5 \longrightarrow (IV)\ +\ C_6H_5\text{-}\overset{O}{\overset{\|}{C}}\text{-}O\bullet \qquad 5.8$$

$$\sim\!\sim CH_2\text{-}\overset{CH_3}{\underset{\underset{OCH_3}{C=O}}{\overset{|}{\underset{|}{C}}}}\bullet\ +\ \bullet C_6H_5 \longrightarrow \sim\!\sim CH_2\text{-}\overset{CH_3}{\underset{\underset{OCH_3}{C=O}}{\overset{|}{\underset{|}{C}}}}\text{-}C_6H_5 \quad (V) \qquad 5.9$$

$$\sim\!\sim CH_2\text{-}\overset{CH_3}{\underset{\underset{OCH_3}{C=O}}{\overset{|}{\underset{|}{C}}}}\bullet\ +\ \bullet CH_2\text{-}\overset{CH_3}{\underset{\underset{OCH_3}{C=O}}{\overset{|}{\underset{|}{C}}}}\text{-}CH_2\sim\!\sim$$

$$\nearrow \sim\!\sim CH_2\text{-}\overset{CH_3}{\underset{\underset{OCH_3}{C=O}}{\overset{|}{\underset{|}{C}}}}\text{-}H\ +\ CH_2{=}\overset{CH_3}{\underset{\underset{OCH_3}{C=O}}{\overset{|}{\underset{|}{C}}}}\text{-}CH_2\sim\!\sim \qquad 5.10$$
$$(VI) \qquad\qquad (VII)$$

$$\searrow \sim\!\sim CH_2\text{-}\overset{CH_3}{\underset{\underset{OCH_3}{C=O}}{\overset{|}{\underset{|}{C}}}}\!\!-\!\!\overset{CH_3}{\underset{\underset{OCH_3}{C=O}}{\overset{|}{\underset{|}{C}}}}\text{-}CH_2\sim\!\sim \qquad 5.11$$
$$(VIII)$$

Structures 5.4–5.11

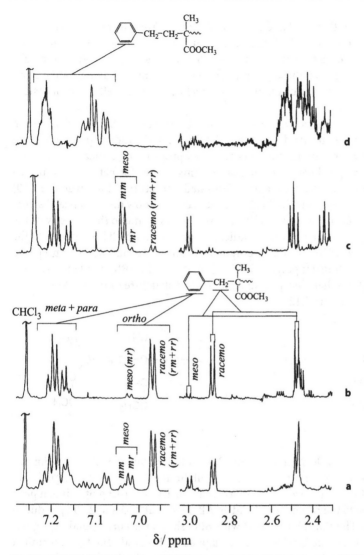

Fig. 5.2. 750 MHz ¹H NMR spectra of PMMA prepared with **a** BPO in toluene at 100 °C, **b** with C₆H₅MgBr in THF at –78 °C, **c** with C₆H₅MgBr in toluene at 40 °C, and **d** with C₆H₅CH₂Li in toluene at –78 °C [6]. 10 wt/vol%,CDCl₃ solution, 55 °C, 45° pulse, pulse repetition time 18 s, 512 scans

the figure. The radically prepared PMMA apparently shows additional signals at 7.05–7.14 ppm, suggesting the presence of another type of phenyl group. The possible source of this is the chain transfer to the solvent, which generated a benzyl radical from toluene to form the end structure III in Structure 5.6. In order to examine this possibility, PMMA having a benzyl group at the initiating chain end was prepared by the polymerization of MMA with benzyllithium. The phenyl-region spectrum of the PMMA, in which the signals are observed in the range from 7.05–7.24 ppm, is shown in Fig. 5.2d. Thus, it is sure that the radically prepared

PMMA contains both phenyl and benzyl end groups. From the relative peak intensities of these two regions, the fractions of structure II in Structure 5.5 and structure III in Structure 5.6 were found to be 75 and 25%, respectively. The results demonstrate that the syntheses of model polymer samples having defined end groups, based on the anionic polymerization technique, is very effective for assigning end group signals.

In the olefinic proton region (Fig. 5.1, range C), signals of the vinylidene group at the chain end are observed at 5.42 and 6.15 ppm. In the radical polymerization of MMA, the termination reaction between two propagating radicals occurs predominantly through disproportionation, leading to the formation of saturated (structure VI in Structure 5.10) and unsaturated (structure VII in Structure 5.10) chain ends. The signals seen in Fig. 5.1, range C clearly show the presence of structure VII. The signal at 5.42 ppm splits into two peaks owing to the tacticity of the terminal diad, whose intensity ratio indicates m/r=26/74, which is very close to the in-chain tacticity (25/75). A model polymer for this end group was also prepared by catalytic chain-transfer polymerization initiated with AIBN in bulk in the presence of cobalt tetraphenylporphyrin at 60 °C. The structure of the PMMA can be depicted as in Structure 5.12:

Structure 5.12

A part of the ^{1}H NMR spectrum of the PMMA thus obtained is shown in Fig. 5.3. In addition to the olefinic proton signals, allylic methylene proton signals are observed at 2.43–2.62 ppm with the expected intensity, consisting of a singlet peak at 2.47 ppm due to the racemo diad and an AB quartet at 2.45 and 2.60 ppm due to the meso diad. The PMMA prepared with BPO shows the corresponding signals as seen in Fig. 5.1, range E, which are overlapped with signals due to the terminal methine proton and benzylic methylene protons as indicated in the figure. As demonstrated by the previous example, the detailed end-group analysis by NMR provides us with important information on the elemental reaction processes in the radical polymerization, such as initiation, termination, and chain-transfer reactions.

As already described, several end-group signals show splittings owing to tacticity near the chain end. The diad tacticity at the initiating chain end having a benzoyloxy group (structure I in Structure 5.4), determined from –OCH$_2$– proton signals (m/r=41/59) (Fig. 5.1, range D) differs from that at the chain end having a phenyl group, determined from ortho proton signals (m/r=29/71) (Figs. 5.1, range B, 5.2a). This indicates that the stereospecificity in the initiation process of radical polymerization is affected by the structure of the initiating radical.

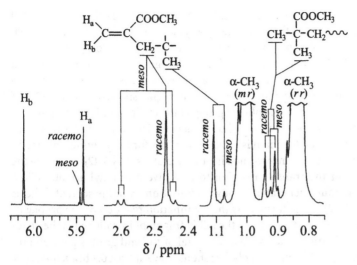

Fig. 5.3. 750 MHz ^{1}H NMR spectrum of PMMA prepared with AIBN in the presence of cobalt tetra-phenylporphyrin in bulk at 60 °C [6]

PMMA is known to undergo unzipping degradation to form MMA monomer starting from the unsaturated chain end (structure VII in Structure 5.10) or a head-to-head linkage formed through a combination reaction (structure VIII in Structure 5.11). When the polymer is prepared in the presence of chain-transfer agents such as thiols, bimolecular termination reactions between propagating PMMA radicals is suppressed by the chain-transfer reaction to thiol (Structure 5.13) and thus the fractions of the polymer molecules containing the unsaturated end group and head-to-head linkage decrease. For example, PMMA prepared

Structure 5.13

with AIBN in the presence of *t*-butyl mercaptan (0.5 mol% of MMA) was found to contain only 3.4% of the vinylidene end group, and its thermal degradation proceeded mostly through main-chain scission and not through the scission of the head-to-head linkage [7] and/or the unzipping from the unsaturated chain end. Consequently, the thermal stability of the PMMA increased greatly. Thus, the analysis of such a small defect in the polymer is sometimes very important to understand its properties.

Solvent fragments are usually incorporated into the polymer chain end through a chain-transfer reaction in radical polymerization. Since radical polymerization of vinyl acetate in aromatic solvents such as benzene is slower than that in ethyl acetate [8], it was suggested that benzene copolymerizes with vinyl acetate, which causes a decrease in the apparent rate of polymerization [9]. Poly(vinyl acetate) prepared with AIBN in benzene showed a small signal in the aromatic region (7.31 ppm), whose ^{1}H T_1 was 0.72 s, which is assigned to the phenyl group attached at the chain end. The intensity measurement of the signal showed that only 15% of the polymer chain contained one phenyl group at the chain end. The possibility of the copolymerization of benzene was thus neglected. The number of AIBN fragments per poly(vinyl acetate) chain is larger than unity for the polymer formed in benzene but that for the polymer formed in ethyl acetate is only 0.42, indicating the higher frequency of the transfer reaction to the solvent in ethyl acetate. When counting the end groups generated through reinitiation from the solvent radicals (Structures 5.14, 5.15), those formed by the termination reaction of the propagating radicals (Structures 5.16, 5.17) and the AIBN fragment introduced at the chain end through the initiation reaction, the sum total of the end groups per polymer molecule amounts to 2.73. Thus the polymer should have at least 0.7 branchings per polymer molecule. The total numbers of end groups in the polymers formed in benzene and chlorobenzene are 1.95 and 1.89, respectively, indicating the polymers have no branching. These results are consistent with the explanation [10, 11] that the propagating radicals in aromatic solvents are stabilized through π-complex formation with the aromatic rings leading to the decrease of the rate of polymerization [12].

Poly(vinyl chloride) usually comprises various types of defects, such as branching, regio-irregular unit, and unsaturated unit, owing to the higher reactivity of the propagating radical [13]. The polymer can be reduced with Bu$_3$SnH to polyethyl-

$$CH_3CH_2OCCH_2 \bullet \; + \; H_2C{=}CH \longrightarrow CH_3CH_2OCCH_2\,CH_2{-}CH \bullet$$

Structure 5.14

Structure 5.15

Structure 5.16

Structure 5.17

ene, the ^{13}C NMR spectrum of which provides structural information on branching and end groups by referring to the spectral data of polyethylene [14–18].

In coordination polymerization of α-olefins such as propylene, the mechanisms of initiation, propagation, and chain-transfer reactions, in addition to the stereospecificity as discussed in Chap. 3, are of prime importance for the understanding of the polymerization mechanism. Two kinds of monomer insertion may be involved; primary insertion and secondary insertion (Structures 5.18, 5.19). These two mechanisms can be distinguished through end-group analysis. For example,

$$M-P + CH_2=CH-CH_3 \quad \xrightarrow{\text{primary insertion}} \quad M-CH_2-CH-P \atop CH_3 \qquad \textbf{5.18}$$

$$M: \text{transition metal} \atop P: \text{polymer chain} \quad \xrightarrow{\text{secondary insertion}} \quad M-CH-CH_2-P \atop CH_3 \qquad \textbf{5.19}$$

Structures 5.18 and 5.19

the ^{13}C NMR spectra of isotactic polypropylene prepared with either racemic ethylenebis(4,5,6,7-tetrahydro-1-indenyl)dichlorozirconium or racemic ethylenediindenyldichlorozirconium and methylalumoxane showed the presence of vinylidene and n-propyl end groups in equal amounts. The presence of the vinylidene end group indicates that the spontaneous chain-transfer process mainly involved β-hydrogen abstraction from a monomer unit at the propagating chain end as shown in Structure 5.18 (Structure 5.20). Initiation of a new polymer chain on a metal–hydrogen bond thus formed proceeds via primary insertion to form the n-propyl end group (Structure 5.21) [19]. If the initiation takes place through secondary insertion, the end group should be an isopropyl end group (Structure 5.22), but this is not the case.

$$M-CH_2-CH-CH_2-CH\sim\!\!\sim \quad \longrightarrow \quad M-H + CH_2=C-CH_2-CH\sim\!\!\sim \atop CH_3 \qquad CH_3 \qquad\qquad\qquad\qquad CH_3 \qquad CH_3$$

142.65 ppm, 109.37

Structure 5.20

$$M-H + CH_2=CH-CH_3 \quad \longrightarrow \quad M-CH_2-CH_2-CH_3 \quad \longrightarrow \quad \longrightarrow \quad \longrightarrow$$

$$\sim\!\!\sim\!\!\sim\!\!\sim CH_2CH_2CH_3$$

37.51   12.38 ppm

Structure 5.21

$$M-H + CH_2=CH-CH_3 \quad \longrightarrow \quad M-CH-CH_3 \atop CH_3 \quad \longrightarrow \quad \longrightarrow \quad \longrightarrow$$

$$\sim\!\!\sim\!\!\sim CH-CH_3 \atop CH_3$$

Structure 5.22

The results indicate that the initiation on a metal hydride bond and the propagation reaction both proceed through primary insertion.

5.2 Analysis of the Polymerization Reaction

As usual organic reactions, polymerization reactions can be followed by NMR spectroscopy by tracing the signal intensity of the monomers. Owing to the time required to start the NMR measurement and to accumulate NMR signals, however, NMR spectroscopy is applicable to relatively slow reactions. Moreover, polymerization reactions often require rigorous conditions; for example, ionic polymerizations should be carried out under dry conditions or under an inert atmosphere, and thus the NMR tubes used in such experiments should be sealed (Sect. 1.4.3)

Polymerization of MMA with t-C$_4$H$_9$MgBr in toluene at low temperature is known to proceed in a living manner and to produce an isotactic PMMA with a narrow molecular-weight distribution. To determine the rate of polymerization, the reaction was carried out in toluene-d_8 at –78 °C in an NMR sample tube sealed under nitrogen. The polymerization was followed by measuring the intensities of the vinylidene methylene proton signals of MMA. The t-C$_4$H$_9$MgBr signal at 1.58 ppm disappeared upon the addition of MMA, indicating a fast initiation reaction. The propagation was relatively slow compared with the initiation reaction, and the rate of polymerization fitted first-order plots; the rate constant for the propagation reaction was found to be 4.3×10^{-4} l mol/s. The fast initiation and slow propagation as well as the living character are responsible for the formation of PMMA with a narrow molecular-weight distribution.

Owing to the high sensitivity of superconducting NMR instruments, it is now possible to observe end-group signals by NMR as described in the previous section. Macromonomers are polymers having polymerizable functions such as vinyl and vinylidene groups at the chain end. Thus polymerization of macromonomers can be followed by tracing the signals due to such end groups. The change in the ^{1}H NMR spectra of a polymerization mixture of an isotactic PMMA macromonomer having a styrene-type end group (Structure 5.23) in toluene-d_8 at 80 °C initiated with AIBN is shown in Fig. 5.4 [20].

Though the accumulation of each spectrum requires 2.5 min, the rate of polymerization is relatively slow and thus the consumption of the macromonomer can be followed easily by ^{1}H NMR spectroscopy. As the polymerization proceeded, the signal intensity of the vinyl group decreased, and the methoxy and t-butyl end

Structure 5.23

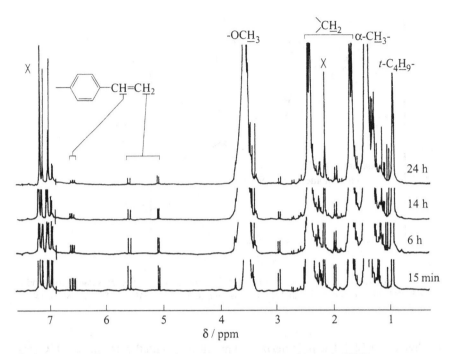

Fig. 5.4. 400 MHz ^{1}H NMR spectra of a polymerization mixture of isotactic PMMA macromonomer ($\bar{M}_n$=2,900) with AIBN in toluene-d_8 at 80 °C [20]. $[M]_0$=0.05 mol/l, $[M]_0/[AIBN]_0$=20 mol/mol. x solvent signals. 30° pulse, pulse repetition time 5 s, 32 scans

group signals became broader. The amount of the macromonomer consumed was determined from the relative intensity of the vinyl methylene proton signals (5.03–5.06 and 5.52–5.56 ppm) to the t-butyl signal (0.90 ppm). The first-order plots of monomer consumption of the isotactic and syndiotactic PMMA macro-monomers at 80 °C are shown in Fig. 5.5. The same information can be obtained by carrying out a series of experiments at different reaction times followed by SEC of polymer samples; in the SEC chromatograms the relative peak areas for poly-macromonomer and the starting macromonomer give the conversion data. However, the NMR method has the advantage that the reaction can be followed using a single reaction mixture with a small amount of sample without scattering of reaction conditions. Moreover, it has been pointed out that the mixture often contains the unimer of the macromonomer, which is formed through the attack of the initiator radical on the macromonomer but failed to propagate further and thus the SEC peak considered to be from the starting macromonomer might contain this product (unimer) that has almost the same molecular size as the macromonomer. By using this NMR technique, the difference in the reactivity of isotactic and syn-diotactic PMMA macromonomers having the same chemical structure was clearly evidenced as shown in Fig. 5.5 [20].

^{31}P NMR was used to identify reactive species in the cationic polymerizations of 2-phenyl-1,2-oxaphospholane with methyl triflate ($CH_3OSO_2CF_3$) and with

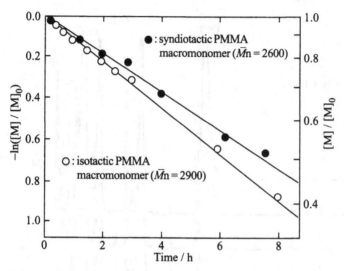

Fig. 5.5. First-order plots of monomer consumption in the polymerization of isotactic and syndiotactic PMMA macromonomers with AIBN in toluene-d_8 at 80 °C [20]. $[M]_0$=0.05 mol/l, $[M]_0/[AIBN]_0$= 20 mol/mol

methyl iodide [21]. The polymerization mixture with methyl triflate at 70 °C gave a ^{31}P NMR signal at 101 ppm due to the phosphonium ion species along with that of the monomer at 109 ppm, suggesting the propagation mechanism shown in Structure 5.24.

Structure 5.24

In contrast, the polymerization mixture with methyl iodide did not show the signal due to a phosphonium-type species but showed a signal at 37 ppm assignable to phosphine oxide with a 3-iodopropyl group as shown in Structure 5.25. The reaction at 0 °C gave the phosphonium-type signal at 101 ppm, which is a precursor to phosphine oxide.

Structure 5.25

In both cases, the concentrations of the active species were constant and thus the rate constant of propagation could be determined from the change of monomer signals, and, interestingly, the covalent-type species (Structure 5.25) exhibited a larger rate constant than the ionic species (Structure 5.24) in this ionic reaction.

Direct and detailed structural analysis of propagating species in the polymerization reaction is often difficult partly owing to the difficulty in sample preparation and stability of the reactive species. Instead of the direct observation, NMR analysis of the terminal structure of the polymer sometimes gives important information on the polymerization mechanism. When the polymerization of alkyl methacrylates is carried out in toluene with a combination of t-C$_4$H$_9$Li and bis(2,6-di-t-butylphenoxy)methylaluminum at low temperature, the reaction proceeds in a living manner and gives heterotactic polymers which comprise an alternating repetition of m and r diads [22, 23]. As depicted from the stereochemical structure of the heterotactic polymer (Structure 5.26), there exist two kinds of propagating anions, r-ended and m-ended anions (Structures 5.27, 5.28), in this polymerization system.

Structure 5.26

Structure 5.27 r-ended anion, $\sim\!\!\sim\!r\,M^-$

Structure 5.28 m-ended anion, $\sim\!\!\sim\!m\,M^-$

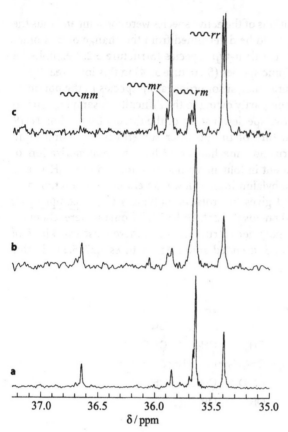

Fig. 5.6. 125 MHz ^{13}C NMR spectra of the methine carbon at the terminating chain end of heterotactic PEMA formed at **a** 28%, **b** 41%, and **c** 100% conversions, respectively [25]

Direct identification of these two types of propagating species is still difficult; however, ^{13}C NMR analysis for the stereochemical sequences at and near the chain end of heterotactic PEMA was found useful to determine the fraction of the two types of species [24, 25]. ^{13}C NMR signals of the terminal methine carbon showed principally four split signals reflecting four possible triad sequences at the chain end, $-mm$, $-mr$, $-rm$, and $-rr$, and were assigned as shown in Fig. 5.6c [25]. The assignments were made by comparing the spectra of three types of stereoregular PEMAs: isotactic, syndiotactic, and heterotactic polymers. The ultimate diads were fixed when the polymerization reaction was quenched with methanol and the chain-end anion was protonated, which reflects the stereoselectivity of the protonation reaction. The m/r ratio of the second diad from the terminal should correspond to the ratio of the m-ended and r-ended anions existing in the polymerization system before quenching the reaction; i.e., the ratio $[\sim\sim\sim rM^-]/[\sim\sim\sim mM^-]$ can be estimated from the terminal triad fractions as

$$[\sim\sim\sim rM^-]/[\sim\sim\sim mM^-] = ([-rm] + [-rr])/([-mm] + [-mr]). \qquad (5.1)$$

The analysis for the polymer obtained at 28% conversion revealed that the ratio was 78/22, i.e., the r-ended anion is more abundant than the m-ended anion, suggesting the higher stability or lower reactivity of the r-ended anion. It is interesting that the spectra of the polymer changed with conversion and became more complicated at high conversion (Fig. 5.6b, c). This is due to the increasing irregularity of the chain-end tacticity at the later stage of the polymerization. Thus, the analysis of the products from the polymerization reaction of interest at different conversion may be useful to understand how the polymerization reaction proceeds.

5.3 Chemical Shifts of Vinyl Monomers and Their Reactivities

Several attempts have been made to correlate the reactivities of vinyl monomers in polymerization or copolymerization with their spectral data [26–38]. Reactivity of vinyl monomers usually depend on the π-electron density on the double bond, particularly on the β carbons. On the other hand, the ^{13}C NMR chemical shift of the β carbon ($\delta_{C\beta}$) also depends on the π-electron density. The higher the π-electron density on the β carbon, the higher the magnetic field where the NMR peak is observed; i.e., as the π-electron density increases, the ^{13}C NMR chemical shift value of the peak becomes smaller. Therefore it should be possible to correlate the magnitude of the shift with the reactivity of the monomer. Systematic studies on the chemical shift of vinyl monomers are important and useful for understanding the mechanisms of polymerization as well as for estimating the reactivity without or prior to performing the experiments.

Radical polymerization and copolymerization can be described by the Alfrey–Price Q– e scheme, where Q represents the intrinsic reactivity of the monomer and e the polarity of the polymer radical and monomer.

A plot of e values of vinyl monomers against their $\delta_{C\beta}$ values is shown in Fig. 5.7 [37]. The e value increases linearly with increasing $\delta_{C\beta}$ value, indicating that the

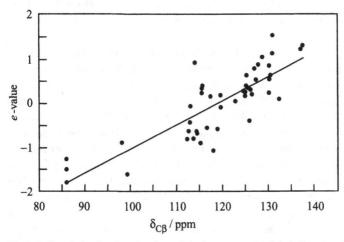

Fig. 5.7. Correlation between e values of vinyl monomers and their $\delta_{C\beta}$ values [37]

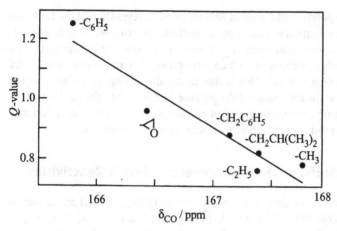

Fig. 5.8. Q value against the chemical shift of the carbonyl carbon (δ_{co}) of various methacrylates [37]

e value increases with decreasing π-electron density on the β carbon of the monomer. This is clear evidence that the e value is a measure of the polarity of the monomer. The correlation can be expressed by Eq. (5.2):

$$e = 0.079\,\delta_{C\beta} - 9.45 . \tag{5.2}$$

Thus the $\delta_{C\beta}$ value can be used as a good guide for the e value.

It is rather difficult to correlate Q values to NMR parameters for a wide range of monomers; however, it is possible for a family of monomers. One example is a plot of Q value against the chemical shift of the carbonyl carbon of various methacrylates (Fig. 5.8). The linear relation in the figure indicates that the π-electron density on the carbonyl carbon increases as Q becomes larger. Owing to the electron-withdrawing character of the carbonyl group, resonance stabilization increases the electron density at the carbonyl carbon. Then, the linear relation should indicate that Q is a measure of the resonance stabilization of the methacrylate monomers [37].

Examination of the relationship between the reactivity of monomers and NMR data is also of significance to the understanding of reactivity of monomers and the mechanism of reaction in ionic polymerization. Plots of $\delta_{C\beta}$ against the relative reactivity, $\log(1/r_1)$, in the copolymerization of MMA with other alkyl methacrylates in toluene by n-C_4H_9Li at $-78\,^\circ C$ are shown in Fig. 5.9. The $\log(1/r_1)$ values increased linearly with increasing $\delta_{C\beta}$, indicating that the lower the π-electron density on the β carbon of the monomer is, the higher the relative reactivities; i.e., the polymerizations are typical anionic polymerizations and the attack of the propagating anion on the β carbon of the monomer is a rate-determining step. The monomers used differ from each other not only in the π-electron density on the β carbon but also in the steric hindrance. Nevertheless all the plots fall on a single line. This indicates that $\delta_{C\beta}$ can be a good measure of the reactivity for a wide range of methacrylate monomers. Similar plots were obtained for the copolymerization

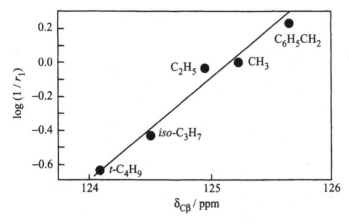

Fig. 5.9. $\delta_{C\beta}$ against the relative reactivity, $\log(1/r_1)$, in the copolymerization of MMA (M_1) with alkyl methacrylates (M_2) in toluene by n-C$_4$H$_9$Li at –78 °C [37]

of methacrylates in THF. The slope of the relationship was similar to that for the polymerization in toluene. So the relative reactivity of alkyl methacrylate in anionic polymerization with n-C$_4$H$_9$Li is controlled mainly by the π-electron density on the β carbon of the monomer, regardless of the bulkiness of the ester group and the solvent [37].

Values of $\delta_{C\beta}$ are also very useful for understanding cationic polymerization. Plots of $\delta_{C\beta}$ against $\log(1/r_1)$ in the cationic copolymerization of styrene with substituted styrenes by SnCl$_4$ at 0 °C are shown in Fig. 5.10. The good linear relation obtained indicates that the relative reactivity is larger for the substituted styrene having the higher π-electron density on the β carbon [35, 37]. These results indicate that the attack of the growing carbonium ion on the β carbon of the monomer is the rate-determining step in this copolymerization. Similar relations between the reactivity of the vinyl monomer and the $\delta_{C\beta}$ value were observed in the cationic

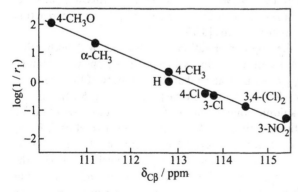

Fig. 5.10. $\delta_{C\beta}$ against $\log(1/r_1)$ in the copolymerization of styrene (M_1) with substituted styrenes (M_2) at 0 °C by SnCl$_4$ [35, 37]

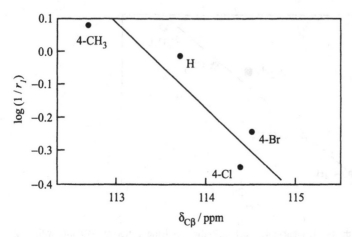

Fig. 5.11. $\delta_{C\beta}$ against $\log(1/r_1)$ in the copolymerization [39] of styrene (M_1) with substituted styrenes (M_2) at 40 °C by TiCl$_3$–Al(C$_2$H$_5$)$_3$ [35, 37]

copolymerization of benzyl vinyl ether with substituted benzyl vinyl ethers and in the copolymerization of phenyl vinyl ether with substituted ones [35, 37].

^{13}C NMR provides us with information on coordinated anionic polymerization. Plots of $\delta_{C\beta}$ against $\log(1/r_1)$ for the copolymerization [39] of styrene with several substituted styrenes by a Ziegler catalyst, TiCl$_3$–Al(C$_2$H$_5$)$_3$, at 40 °C are shown in Fig. 5.11 [35, 37]. The linear relation in the figure is a typical feature of ordinary cationic polymerization; the higher the π-electron density on the β carbon, the higher the reactivity of the monomer. Natta and coworkers [39] reported that the coordination of olefinic monomer to the more electron-deficient part of the catalyst complex is the fundamental stage in the stereospecific polymerization by a Ziegler catalyst. The results obtained here are evidence that the coordination of the monomer to Ti or Al in the catalyst is the rate-determining step and are also clear indication that the vinyl group of the monomer is bound to the catalyst complex at its β position in the coordination step. Similar results were obtained in the copolymerizations [40] of styrene as M_1 with α-olefins by TiCl$_3$–Al(C$_2$H$_5$)$_3$ at 40 °C, indicating the importance of the coordination process in the propagation reaction with titanium-containing initiator systems [35, 37].

On the other hand, in the copolymerization of 1-butene (M_1) with other olefins by VCl$_3$–Al(C$_6$H$_{13}$)$_3$ [41] a linear relation with a positive slope was obtained between $\log(1/r_1)$ and the $\delta_{C\beta}$ value, indicating that the reactivity of the monomer increased with decreasing π-electron density on the β carbon (Fig. 5.12). This is a typical feature of ordinary anionic polymerization. Similar results were obtained [37] for the copolymerization of styrene (M_1) with several α-olefins (M_2) by VCl$_3$–Al(i-C$_4$H$_9$)$_3$ [42] or of 1-butene (M_1) with several α-olefins (M_2) by VCl$_4$–Al(i-C$_6$H$_{13}$)$_3$ [41]. These results indicate that the polymerizations with the vanadium-containing initiators proceed through a mechanism similar to that of usual anionic polymerization and that the coordination is less important in the propagation reaction [37].

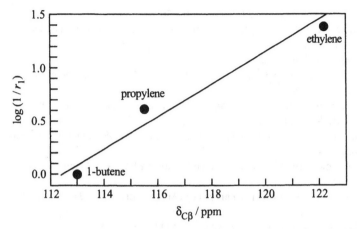

Fig. 5.12. $\delta_{C\beta}$ against $\log(1/r_1)$ in the copolymerization of 1-butene (M_1) with α-olefins by VCl_3–$Al(C_6H_{13})$ [37]

^{1}H NMR chemical shifts are affected by the π-electron density of the attached carbon but also by other factors, such as diamagnetic anisotropy. Moreover, the chemical shift range for ^{1}H NMR is much smaller than that for ^{13}C NMR. Thus it would not be so appropriate to use the ^{1}H NMR chemical shift as a measure of the reactivity of a vinyl monomer. However, it was found that the chemical shifts of β-methylene protons, particularly of the proton trans to the substituent of the vinyl monomer could be used as a good practical measure of its relative reactivity. The correlation between the reactivity and the chemical shift is closer in the case where the relative reactivities among the homologues of monomers are studied [37, 38]. ^{1}H NMR spectra can be measured much easily for a short time compared with the corresponding ^{13}C NMR spectra. This is an advantage in using ^{1}H NMR spectra as a measure of reactivity.

References

1. BEVINGTON JC, EBDON JR, HUCKERBY TN (1985) Eur Polym J 21:685
2. AXELSON DE, RUSSELL KE (1985) Prog Polym Sci 11:221
3. BEVINGTON JC, MELVILLE HW, TAYLOR RP (1954) J Polym Sci 12:449
4. BEVINGTON JC, MELVILLE HW, TAYLOR RP (1954) J Polym Sci 14:463
5. STARNES WH JR, PLITZ IM, SCHILLING FC, VILLACORTA GM, PARK GS, SAREMI AH (1984) Macromolecules 17:2507
6. HATADA K, KITAYAMA T, UTE K, TERAWAKI Y, YANAGIDA T (1997) Macromolecules 30:6754
7. KASHIWAGI T, INABA A, BROWN JE, HATADA K, KITAYAMA T, MASUDA E (1986) Macromolecules 19:2160
8. BURNETT GM, MELVILLE GM (1947) Discuss Faraday Soc 2:322
9. STOCKMAYER WH, PEEBLES LH JR (1953) J Am Chem Soc 75:2278
10. KAMACHI M, LIAW DJ, NOZAKURA S (1979) Polym J 11:921

11. KAMACHI M, SATOH J, LIAW DJ (1979) Polym Bull 1:581
12. HATADA K, TERAWAKI Y, KITAYAMA T, KAMACHI M, TAMAKI M (1981) Polym Bull 4:451
13. GUYOT A (1985) Pure Appl Chem 57:833
14. STARNES WH JR (1985) Pure Appl Chem 57:1001
15. STARNES WH JR, VILLACORTA GM, SCHILLING FC, PLITZ IM, PARK GS, SAREMI AH (1985) Macromolecules 18:1780
16. STARNES WH JR, SCHILLING FC, PLITZ IM, CAIS RE, FREED DJ, HARTLESS RL, BOVEY FA (1983) Macromolecules 16:790
17. LLAURO-DARRICADES MF, BENSEMRA N, GUYOT A, PETIAUD R (1989) Macromol Chem Macromol Symp 29:171
18. STARNES WH JR, SCHILLING FC, ABBAS KB, CAIS RE, BOVEY FA (1979) Macromolecules 12:556
19. GRASSI A, ZAMBELLI A, RESCONI L, ALBIZZATI E, MAZZOCCHI R (1988) Macromolecules 21:617
20. MASUDA E, KISHIRO S, KITAYAMA T, HATADA K (1991) Polym J 23:847
21. KOBAYASHI S, SUZUKI M, SAEGUSA T (1984) Macromolecules 17:107
22. KITAYAMA T, ZHANG Y, HATADA K (1994) Polym J 26:868
23. KITAYAMA T, ZHANG Y, HATADA K (1994) Polym Bull 32:439
24. KITAYAMA T, HIRANO T, HATADA K (1996) Polym J 28:61
25. KITAYAMA T, HIRANO T, HATADA K (1996) Polym J 28:1110
26. ITO T, OTSU T, IMOTO M (1996) J Polym Sci Part B Polym Phys 4:81
27. (a) SEMCHIKOV YD, EGOROCHKIN AN, RYABOV AV (1973) Vysokomol. Soedin Ser B 15:993; (b) Chem Abstr 81:64049c
28. FUJIHARA H, MATSUZAKI K, MATSUBARA Y, ISHIHARA M, MAESHIMA T (1979) J Macromol Sci Chem 13:1081
29. (a) VORONOV SA, PUCHIN VA, KOSIK LA, TOKAREV VS, KISILEV EM (1978) Vysokomol. Soedin Ser B 20:577; (b) Chem Abstr 89:164033b
30. HIGASHIMURA T, OKAMURA S, MORISHIMA I, YONEZAWA T (1969) Polym Lett 7:23
31. HATADA K, NAGATA K, YUKI H (1970) Bull Chem Soc Jpn 43:3267
32. HERMAN JJ, TEYSSIÉ P (1978) Macromolecules 11:839
33. BOREHARDT JK, DALRYMPLE ED (1982) J Polym Sci Polym Chem Ed 20:1745
34. BORCHARDT JK (1985) J Macromol Sci Chem A22:1711
35. HATADA K, NAGATA K, HASEGAWA T, YUKI H (1977) Makromol Chem 178:2413
36. JENKINS AD, HATADA K, KITAYAMA T, NISHIURA T (2000) J Polym Sci Part A Polym Chem 38:4336
37. HATADA K, KITAYAMA T, NISHIURA T, SHIBUYA W (2002) Curr Org Chem 16:121
38. HATADA K, KITAYAMA T, NISHIURA T, SHIBUYA W (2002) J Polym Sci Part A Polym Chem 40:2134
39. NATTA G, DANUSSO F, SIANESI D (1959) Makromol Chem 30:238
40. ANDERSON IH, BURNETT GM, GEDDES WC (1967) Eur Polym J 3:16
41. NATTA G, VALVASSORI A, SARTORI G (1974) In: Kennedy JP, Tornqvist EGM (eds) Polymer chemistry of synthetic elastomers. Wiley-Interscience, New York, p 597
42. BAKER B, TAIT PJT (1967) Polymer 8:225

6 Two-dimensional NMR Spectroscopy

6.1 Principles of 2D NMR

Two-dimensional NMR has become an important class of NMR experiments in elucidating the structures of organic and polymeric compounds. In the typical NMR experiments that we have discussed in the preceding chapters, the frequency-domain spectra are obtained in one-dimension of frequency after Fourier transforming the FID signals obtained in one dimensional time (acquisition time or detection time). In contrast, 2D NMR experiments involve another variable time dimension, the so-called evolution time as seen in Fig. 6.1 [1]. By changing the evolution time t_1, a set of FID data are acquired along the detection time t_2 (Fig. 6.1) [1]. Fourier transformation in the t_2 axis gives a set of NMR data in the F2 frequency axis along with the t_1 axis (Fig. 6.2a) [2]. The figure demonstrates an oscillation of peak intensity along with the t_1 axis (Fig. 6.2b) [2]. The Fourier transformation in the t_1 axis gives rise to another frequency axis, F1. Thus, the NMR data set is Fourier-transformed in the two time axes, t_2 and t_1, which produce the 2D NMR data in two frequency axes, F1 and F2 (Fig. 6.3) [2]. The 2D NMR data are displayed either in a stacked plot (Fig. 6.3a) [2] or in a contour plot (Fig. 6.3b) [2]. Depending on the pulse sequences used, different types of 2D NMR experiments can be executed, such as shift correlation spectroscopy and J-resolved spectroscopy, both of which rely on spin coupling, and NOE spectroscopy, which utilizes through-space dipolar interaction.

When the pulses in Fig. 6.1 are 90°, the 2D NMR is so-called COSY, in which spin-coupled homonuclear spins show "cross-peaks" in addition to the peaks of their own that appear on the diagonal, "diagonal peaks" (Fig. 6.4) [1]. Thus, by tracing the cross-peaks, one can find the spin-coupling networks that lead to the elucidation of molecular structures of the samples. In this case, the two axes of frequency are the same. When the two frequency axes represent different nuclei, such as ^1H and ^{13}C, the method is called heteronuclear chemical shift correlation (HETCOR) spectroscopy.

Another class of 2D NMR is referred to as J-resolved spectroscopy, or simply J-spectroscopy, in which one axis represents the chemical shift and the other the J values of the spin couplings. In addition to spin-coupling interaction through the connecting bonds between the interacting nuclei, through-space interaction, such as dipolar coupling, which gives rise to the NOE can also be used in 2D NMR. The experiment is called NOESY, in which two nuclei existing in proximity show a

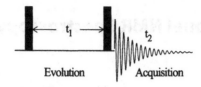

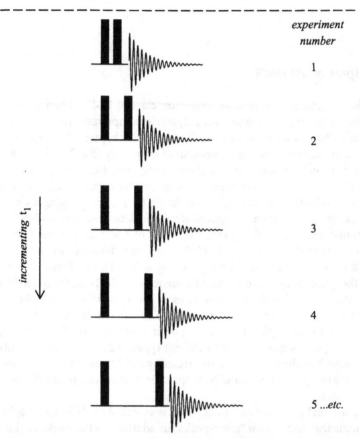

Fig. 6.1. A generalized scheme for the collection of a 2D NMR data set. The experiment is repeated many times with the t_1 period incremented at each stage and the resulting FIDs stored separately. Following double Fourier transformation with respect first to t_2 and then to t_1, the 2D spectrum results [1]

cross-peak and is effective to study molecular structure including the conformation. Several variants of 2D NMR, some of which are explained in detail in the following sections with examples, are listed in Table 6.1 [1].

In 2D NMR experiments, a set of FID data is acquired by changing the evolution time t_1. The number of FIDs in the data set corresponds to the number of data point in the F1 axis that limits the digital resolution. Owing to the available time for the measurement, however, the numbers are usually 2^7–2^8 (128–256), much smaller than that of conventional one-dimensional spectra, and thus the

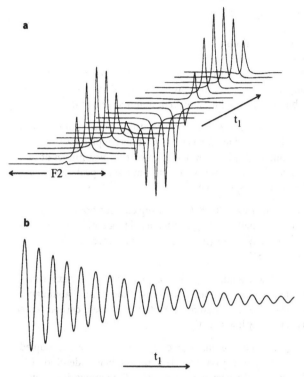

Fig. 6.2. **a** Amplitude modulation of a singlet resonance as a function of the evolution period t_1 and **b** an FID (interferogram) for the t_1 domain produced by the variation in the peak intensity of the resonance [2]

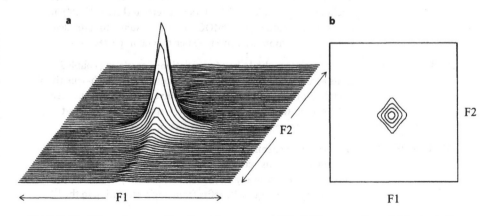

Fig. 6.3. The 2D spectrum resulting from the sequence of Fig. 6.1 for a sample containing a single uncoupled spin: **a** stacked plot, **b** contour plot [2]

Table 6.1. Principal 2D NMR techniques [1]

Technique	Principal applications
COSY	Correlation spectroscopy: Correlating coupled homonuclear spins. Typically used for correlating protons coupled over two or three bonds, but may be used for any high-abundance nuclide
DQF-COSY	Double-quantum filtered COSY: Correlating coupled homonuclear spins. Typically used for correlating protons coupled over two or three bonds. Additional information on the magnitudes of the coupling constants may be extracted from the 2D peak fine structure. Singlets are suppressed
Long-range enhanced COSY	Long-range enhanced COSY: Correlating coupled homonuclear spins through small couplings. Often used to identify proton correlations over many bonds (more than three); hence, also known as long-range COSY
INADEQUATE	Incredible natural abundance double quantum transfer experiment: Correlating coupled homonuclear spins of low natural abundance (below 20%). Typically used for correlating adjacent carbon centers, but has extremely low sensitivity
HMQC	Heteronuclear multiple-quantum correlation: Correlating coupled heteronuclear spins across a single bond and hence identifying directly connected nuclei, most often ^1H–^{13}C. Employs detection of high-sensitivity nuclides, for example, ^1H, ^{19}F, ^{31}P (an "inverse technique")
HSQC	Heteronuclear single-quantum correlation: Correlating coupled heteronuclear spins across a single bond and hence identifying directly connected nuclei. Employs detection of high-sensitivity nuclides, for example, ^1H, ^{19}F, ^{31}P (an "inverse technique"). Provides improved resolution over HMQC so is better suited for crowded spectra but is more sensitive to experimental imperfections
HMBC	Heteronuclear multiple-bond correlation: Correlating coupled spins across multiple bonds. Employs detection of high-sensitivity nuclides, for example, ^1H, ^{19}F, ^{31}P (an "inverse technique"). Essentially HMQC tuned for the detection of small couplings. Most valuable in correlating ^1H–^{13}C over two or three bonds
^{13}C–^1H HETCOR	Heteronuclear chemical shift correlation: Correlating coupled heteronuclear spins across a single bond. Employs detection of the lower-γ nuclide, typically ^{13}C, so has significantly lower sensitivity than inverse techniques. Benefits from high-resolution in the ^{13}C dimension

Table 6.1 (continued)

Technique	Principal applications
Heteronuclear *J*-resolved	Separation of heteronuclear couplings (usually ^{1}H–X) from chemical shifts. Used to determine the multiplicity of the heteroatom or to provide direct measurement of heteronuclear coupling constants
Homonuclear *J*-resolved	Separation of homonuclear couplings (usually ^{1}H–^{1}H) from chemical shifts. Used to provide direct measurement of homonuclear coupling constants or to display resonance chemical shifts without homonuclear coupling fine-structure (e.g., "proton-decoupled" proton spectra)
NOESY	2D nuclear Overhauser and exchange spectroscopy: Establishing NOEs and hence spatial proximity between protons. Suitable for "small" (M_r<1,000) and "large" molecules (M_r>2000) for which NOEs are positive and negative respectively, but may fail for mid-sized molecules (zero NOE). Estimates of internuclear separations can be obtained in favorable cases. Moreover, the NOESY cross-peaks sharing the same phase as the diagonal appear between the protons that exchange chemically with each other.

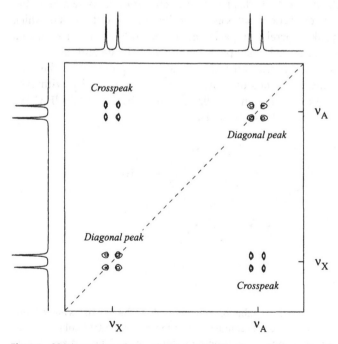

Fig. 6.4. COSY spectrum of a coupled, two-spin AX system with resonance frequencies v_A and v_X. The diagonal peaks are equivalent to those observed in the 1D spectrum, while the cross-peaks provide evidence of coupling between the correlated spins [1]

data are zero-filled to increase the data points before 2D Fourier transformation.[1] The number of scans required to obtain each FID is the critical parameter to determine how long the measurement takes to be completed.

Window functions used in the manipulation of 2D NMR data are not only of a simple exponential form but are rather special ones, a sine function, a sine-squared function, and so on, and are contained in NMR software. A sine function and a sine–bell function are often used in ^{1}H COSY, and an exponential function is used in J-resolved spectroscopy and ^{13}C–^{1}H HETCOR. For a more complete description of 2D NMR, the reader may refer to standard textbooks [1–3].

6.2 ^{1}H COSY (2D Correlation Spectroscopy)

It has become feasible to assign an NMR spectrum "one after another" by using 2D NMR spectroscopy, and thus 2D NMR is effectively employed for structure determinations of various types of compounds. Recently, 2D NMR has been used in the field of polymer chemistry to make structural analysis of polymers and oligomers, assignments of the signals sensitive to stereochemical sequences (tacticity) of vinyl polymers and monomer sequences of copolymers (see also Sect. 3.2) [4–7]. The basic principles of 2D NMR analysis of stereochemical assignments for vinyl-type polymers and several examples are explained in Sect. 3.2.

A ^{1}H COSY spectrum is usually displayed as a contour or cross-sectional diagram with the same spectral coordinates as the ordinate and the abscissa in which the so-called cross-peaks (correlation peaks) appear in addition to diagonal own peaks. A cross-peak is observed at the point where the ordinate and the abscissa cross each other; the coordinates of the cross-peak correspond to the chemical shifts of signals for the two kinds of protons that are spin-coupled with each other. So a close inspection of the spin-coupling network observed in the ^{1}H COSY spectrum is very useful for peak assignments.

Structure 6.1   MMA *racemo* dimer

---

[1] When it is difficult to increase the number of data points along the F1 axis owing to limited accumulation time, it is necessary to compensate for the lack of digital resolution in the data processing. The most popular data processing for such a purpose is zero-filling, in which data points with zero intensity are added at the tail of the FID before Fourier transformation.

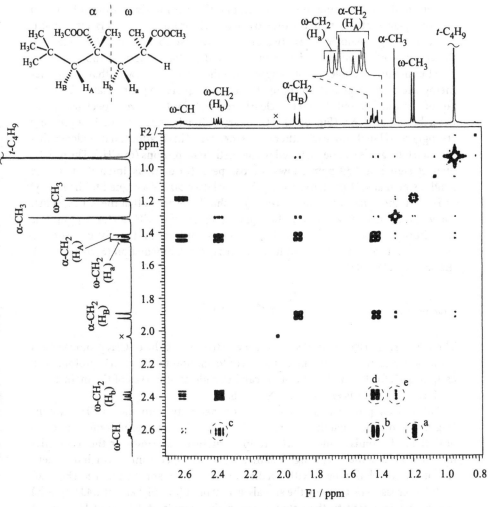

Fig. 6.5. 500 MHz ¹H COSY spectrum of the racemo dimer of MMA with a t-C_4H_9 end group measured in nitrobenzene-d_5 at 90 °C [8, 9]. The peak labeled x is due to a small amount of water. Recycle time 5 s, 16 scans, spectral width 1,585.8 Hz, data points 2,048 (F2), 256 (F1)

The ¹H COSY spectrum of a racemo dimer (Structure 6.1) of MMA prepared with t-C_4H_9MgBr in toluene and subsequently isolated by HPLC separation is shown in Fig. 6.5 [8, 9]. In the COSY spectrum the ordinary NMR signals of the dimer are shown on the ordinate and the abscissa as mentioned previously.

It is clear that the multiplet signal at 2.62 ppm can be assigned to the methine proton at the end of the chain (ω-CH) from the chemical shift and the splitting pattern. The ω-CH signal shows cross-peaks, with signals at 1.19 (a), 1.43 (b), and 2.39 (c) ppm. This indicates that the doublet at 1.19 ppm is due to the methyl protons in the ω-monomer unit (ω-CH$_3$) and that the multiplet around 1.43 ppm and the quartet at 2.39 ppm are due to the ω-CH$_2$ protons (H$_a$ and H$_b$; the definitive peak assignments for these two protons are described in Sect. 6.6). The methylene proton signal around 1.43 ppm should be the overlap of the quartet at 1.434 ppm and a doublet at 1.426 ppm since the intensity is twice that of another methylene proton signal at 2.39 ppm; the doublet signal is explained later. The overlap of the quartet and doublet is clearly revealed by J-resolved spectroscopy as described in Sect. 6.8 (Fig. 6.16). The quartet signals at 1.434 and 2.39 ppm are strongly correlated with each other (cross-peak d). This is additional evidence that the quartet signals can be assigned to the methylene protons (ω-CH$_2$). The ω-CH$_2$ quartet signal at 2.39 ppm shows a cross peak (e) of weak intensity with the singlet signal at 1.31 ppm, indicating that the latter can be assigned to the methyl protons in the α-monomer unit (α-CH$_3$). It should be noted that the α-CH$_3$ signal shows no connectivity with another quartet signal of ω-CH$_2$ at 1.434 ppm. It is well known that the effective long-range coupling in the fully saturated H–C–C–C–H system is confined to a planar zigzag conformation (Structure 6.2), the so-called W rule.

Structure 6.2

Then, the previously mentioned results mean that one of the ω-CH$_2$ protons which shows a signal at 2.39 ppm adopts a W conformation to the α-CH$_3$ proton. Peak assignments for H$_a$ and H$_b$ of ω-CH$_2$ can be made on the basis of the main-chain conformation. This is explained in Sect. 6.6.

The ω-CH$_3$ proton signal shows no connectivity with the ω-CH$_2$ protons (Fig. 6.5), indicating that the ω-CH$_3$–ω-CH bond adopts a gauche conformation to both the C–H bonds of the ω-CH$_2$ group (Structure 6.6). However, the connectivity is observed in the long-range enhanced ^{1}H COSY spectrum in which the intensity of the correlation peak with smaller coupling constant is enhanced (Fig. 6.6) [10]. The cross-peak between the signals at 1.19 ppm (ω-CH$_3$) and at 1.434 ppm (a) is stronger in intensity than that between the signals at 1.19 (ω-CH$_3$) and at 2.39 ppm (b). This indicates that one of the ω-CH$_2$ protons showing signals at 1.434 ppm adopts a conformation approaching the W conformation with the ω-CH$_3$ proton (Structure 6.7). Further discussion on the conformation of the whole molecule is made in Sect. 6.6.

The α-CH$_3$ singlet signal shows correlation peaks with the doublets at 1.426 (c) and 1.90 (d) ppm. This pair of doublets is assigned to the α-CH$_2$ protons; the two doublets show connectivity (cross-peak e) with each other as depicted in the figure. Nonequivalence of these two methylene protons is due to the fact that they are adjacent to the chiral center of the quaternary carbon. The cross-peak between the α-CH$_3$ signal and the α-CH$_2$ signal at 1.90 ppm (d) is stronger than that between the α-CH$_3$ signal and the α-CH$_2$ signal at 1.426 ppm (c). This indicates that one of the α-CH$_2$ protons showing a signal at 1.90 ppm adopts the W conformation to α-CH$_3$.

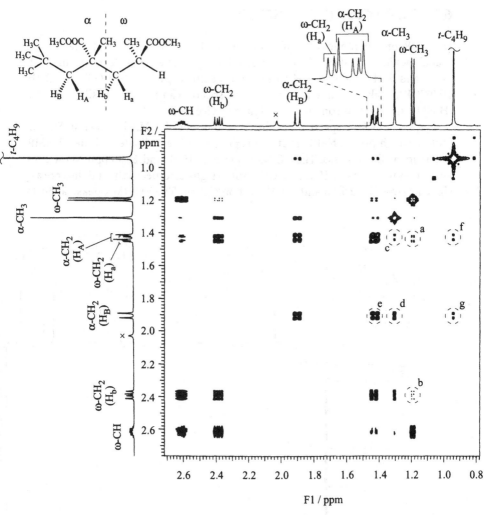

Fig. 6.6. Long-range enhanced 500 MHz ¹H COSY spectrum of the racemo dimer of MMA measured in nitrobenzene-d_5 at 90 °C [10]. The peak labeled x is due to a small amount of water. Recycle time 6 s, 32 scans, spectral width 1,581.7 Hz, 2,048 data points ($F2$), 256 data points ($F1$). A delay time of 0.2 s was adopted for long-range enhancement

The singlet signal at 0.94 ppm shows cross-peaks with the two α-CH₂ doublet signals in Figs. 6.5 and 6.6, and is assigned to t-butyl protons accordingly. In this case both α-CH₂ protons (H$_A$ and H$_B$; the peak assignments for these protons are described in Sect. 6.6) can adopt the W conformation to the t-butyl proton. The two cross-peaks are enhanced in intensity in the long-range enhanced spectrum (Fig. 6.6; cross-peaks f, g).

6.3 ^{13}C–^{1}H COSY (^{13}C–^{1}H HETCOR)

HETCOR or heteronuclear shift correlation (HSC) is the 2D NMR technique that allows identification of coupling relationships in heteronuclear systems such as ^{13}C and ^{1}H. By using ^{13}C–^{1}H correlation spectroscopy (^{13}C–^{1}H COSY or ^{13}C–^{1}H HETCOR), peak assignments for ^{13}C NMR spectra can be made easily when the ^{1}H NMR spectrum is completely assigned or vice versa.

The ^{13}C–^{1}H COSY spectrum of the racemo dimer of MMA is shown in Fig. 6.7 [10], in which the ^{13}C NMR spectrum is plotted as the ordinate and the ^{1}H NMR spectrum as the abscissa. The ^{13}C NMR peaks at 19.80 and 21.96 ppm show connectivities with the ω-CH$_3$ and α-CH$_3$ proton signals, respectively, and thus are assigned to ω-CH$_3$ carbon and α-CH$_3$ carbon signals. The ^{13}C NMR peaks at 30.84

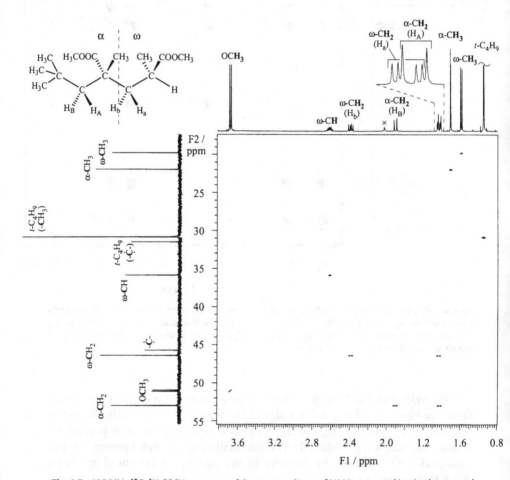

Fig. 6.7. 125 MHz ^{13}C–^{1}H COSY spectrum of the racemo dimer of MMA measured in nitrobenzene-d_5 at 90 °C [10]. Recycle time 3 s, 128 scans. F2 spectral width 4,783.0 Hz (^{13}C), 2,048 data points. F1 spectral width 1,512.2 Hz (^{1}H), 256 data points. J_{C-H} 140 Hz

and 35.86 ppm are assigned similarly to methyl carbons of the *t*-butyl group and the ω-CH carbon, respectively. The peak at 31.43 ppm is assigned to the quaternary carbon of the *t*-butyl group since it shows no cross-peaks and is observed in the shift range of aliphatic carbons. The peak at 45.72 ppm showing no cross-peaks can be ascribed to the quaternary carbon of the main chain, the lower shift of which is due to the electron-withdrawing $-COOCH_3$ substituent. The definite assignments for these quaternary carbons are described in Sects. 6.4 and 6.6. The peak at 46.39 ppm has cross-peaks with two ω-CH_2 proton signals and is assigned to the ω-CH_2 carbon. A ^{13}C NMR peak that shows two cross-peaks with ^1H NMR signals is a typical pattern for the methylene carbon having nonequivalent protons. The peak at 52.98 ppm is assigned to the α-CH_2 carbon, similarly to the case of the ω-CH_2 carbon. The residual two ^{13}C peaks at 50.96 and 51.10 ppm are assigned to two OCH_3 carbons, respectively. Two carbonyl carbon peaks are observed at 176.71 and 177.20 ppm that are not shown in the figure. The peak assignments for the OCH_3 and C=O carbons are discussed in Sect. 6.4. Heteronuclear single-quantum correlation (HSQC) and HMQC (heteronuclear multiquantum correlation) spectroscopies give almost identical information as ^{13}C-^1H COSY and have been used more frequently than the COSY owing to their higher sensitivity (Table 6.1).

6.4 Heteronuclear Multiple-Band Correlation Spectroscopy

2D spectroscopic correlations can also be established between carbons and neighboring protons connected through two or three bonds. The experimental technique is commonly called proton-detected HMBC spectroscopy and involves proton–carbon connectivities through couplings over two or three bonds ($^nJ_{CH}$, n=2,3).[2] The technique can be particularly valuable for the peak assignments of nonprotonated carbons, such as quaternary and carbonyl carbons. The HMBC spectrum of the racemo dimer of MMA in Fig. 6.8 shows the correlations between CH_2 and C=O carbons [10]. The C=O carbon signal at 177.20 ppm shows correlations to α-CH_2 and ω-CH_2 proton signals, and the C=O signal at 176.71 ppm only shows correlation to ω-CH_2 proton signals; thus, the signal at 177.20 ppm can be assigned to the C=O carbon of the ester group in the α-monomer unit (α-C=O) and that at 176.71 ppm to the ω-C=O carbon.

---

[2] The HMBC experiment is one of the commonly employed inverse (or indirect) detected heteronuclear correlation spectroscopies. In the inverse detection experiments, for example in the case of ^1H-^{13}C heteronuclear correlation spectroscopy, the abundant nuclei (^1H) are first excited, and the magnetization is then transferred to the less abundant and low-γ nuclei (^{13}C). After an evolution period, the magnetization of the ^{13}C nuclei is again transferred to the high-γ nuclei (^1H) to obtain FID under broadband decoupling of ^{13}C nuclei to give signals free from ^1H-^{13}C couplings. The inverse detection provides higher sensitivity than the traditional ^{13}C-^1H COSY. A similar inverse detected heteronuclear correlation spectroscopy termed HSQC gives principally the same results with enhanced resolution but with lower sensitivity. For detailed descriptions, refer to Ref. [1].

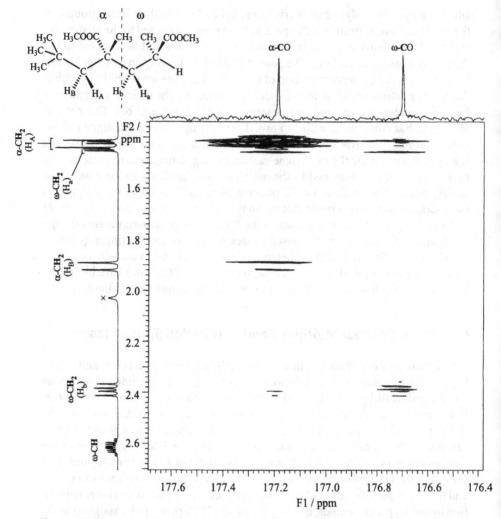

Fig. 6.8. HMBC spectrum (I) of the racemo dimer of MMA measured in nitrobenzene-d_5 at 90 °C – correlation between CH$_2$ and CO [10]. Recycle time 4 s, 24 scans. F2 spectral width 1,555.3 Hz (^{1}H), 1,024 data points. F1 spectral width 21,025.0 Hz (^{13}C), 512 data points. $^1J_{C-H}$ 140 Hz, $^nJ_{C-H}$ 8.8 Hz

Peak assignments for OCH$_3$ proton and carbon signals cannot be made by ^{1}H COSY and ^{13}C–^{1}H COSY techniques since ^{1}H–^{1}H couplings are absent as to the OCH$_3$ signals. The HMBC spectrum provides valuable connectivity for the assignments as shown in Fig. 6.9 [10], which shows the correlations between the α-CO signal at 177.20 ppm and the OCH$_3$ proton signal at 3.672 ppm and between the ω-CO carbon signal at 176.71 ppm and the OCH$_3$ proton signal at 3.688 ppm. The results indicate that the signals at 3.672 and 3.688 ppm can be ascribed to α-OCH$_3$ and ω-OCH$_3$ protons, respectively. Then, the OCH$_3$ carbon signals at 50.96 and

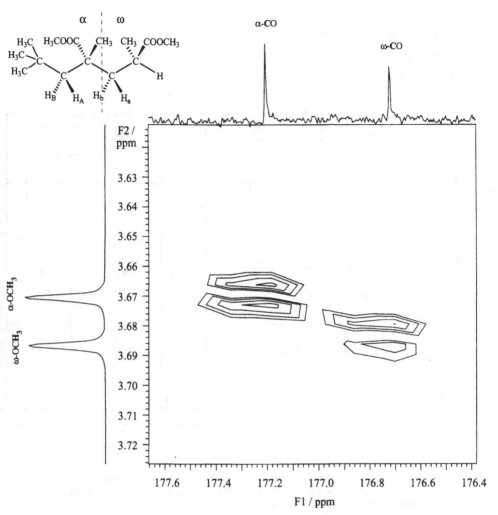

Fig. 6.9. HMBC spectrum (II) of the racemo dimer of MMA measured in nitrobenzene-d_5 at 90 °C – correlation between OCH$_3$ and CO [10]. Recycle time 3 s, 32 scans. *F2* spectral width 1,711.0 Hz (^{1}H), 1,024 data points. *F1* spectral width 20,570.8 Hz (^{13}C), 512 data points. $^1J_{C-H}$ 140 Hz, $^nJ_{C-H}$ 8.8 Hz

51.10 ppm can be assigned to α-OCH$_3$ and ω-OCH$_3$ carbons, respectively, by using the ^{13}C–^{1}H COSY spectrum.

The ^{13}C NMR peaks at 31.43 and 45.72 ppm of the racemo dimer are the signals which show no correlation peaks in the aliphatic carbon region of the ^{13}C–^{1}H COSY spectrum. In the HMBC spectrum (Fig. 6.10) [10] the former peak shows connectivity with α-CH$_2$ proton signals and the latter with both α-CH$_2$ and ω-CH$_2$ proton signals. This clearly indicates that the peak at 31.43 ppm can be assigned to the quaternary carbon of the t-C$_4$H$_9$ group and that at 45.72 ppm to the quaternary carbon in the chain.

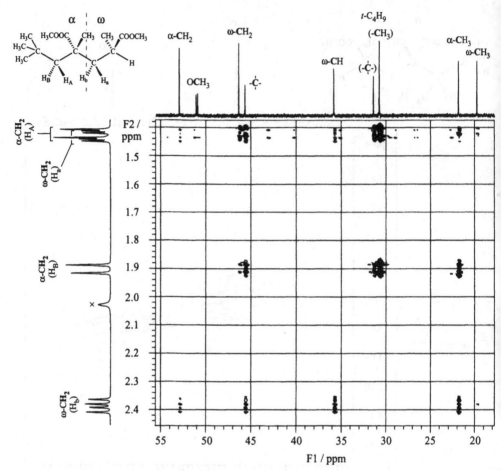

Fig. 6.10. HMBC spectrum (III) of the racemo dimer of MMA measured in nitrobenzene-d_5 at 90 °C [10]. Recycle time 3 s, 24 scans. *F2* spectral width 1,711.0 Hz (^{1}H), 1,024 data points. *F1* spectral width 20,570.8 Hz (^{1}H), 512 data points. $^1J_{C-H}$ 140 Hz, $^nJ_{C-H}$ 8.8 Hz

6.5　Two-dimensional Nuclear Overhauser Enhancement Spectroscopy

The NOESY technique is concerned with the direct and through-space magnetic interactions (dipolar couplings) that give rise to the NOE. NOESY gives the cross-peaks that indicate NOE interactions between the correlated spins, i.e., spatial proximity between the spins (at most about 4 Å).

The 2D NOESY spectrum of the racemo dimer of MMA in nitrobenzene-d_5 at 90 °C is shown in Fig. 6.11 [10]. The *t*-butyl proton signal at 0.94 ppm shows cross-peaks with α-CH$_2$ signals at 1.426 (a) and 1.90 (b) ppm and with α-CH$_3$ signal at 1.31 ppm (c), indicating that the *t*-butyl group is in the proximity of α-CH$_2$ and

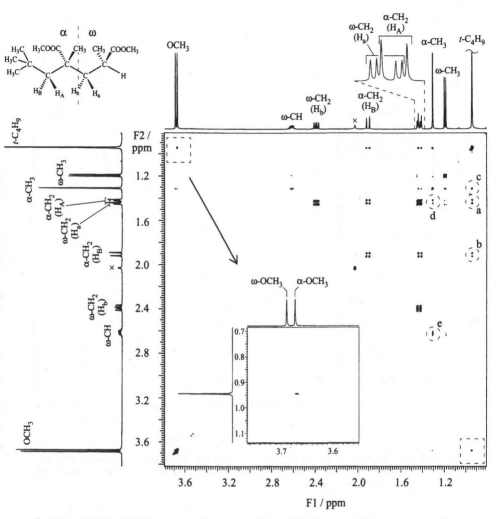

Fig. 6.11. 500 MHz NOESY spectrum of the racemo dimer of MMA measured in nitrobenzene-d_5 at 90 °C [10]. Recycle time 5 s, 16 scans, spectral width 1,640.1 Hz, 2,048 data points (*F2*), 256 data points (*F1*), mixing time 2.28 s

α-CH₃ groups. The *t*-butyl proton signal also shows a cross-peak with the α-OCH₃ signal at 3.672 ppm but not with ω-OCH₃ signal at 3.688 ppm (see insert in Fig. 6.11). The spatial distances between the OCH₃ groups and the *t*-butyl group depend on the main-chain conformation. The NOESY correlations between *t*-butyl and OCH₃ protons indicate that the main-chain conformation is close to a zigzag one. The α-CH₃ signal is correlated with one of the α-CH₂ proton signals at 1.426 ppm (d) and with the ω-CH signal (e) as shown in Fig. 6.11. The phenomena can also be ascribed to the main-chain conformation. A detailed discussion on the conformation of the racemo dimer is given in Sect. 6.6.

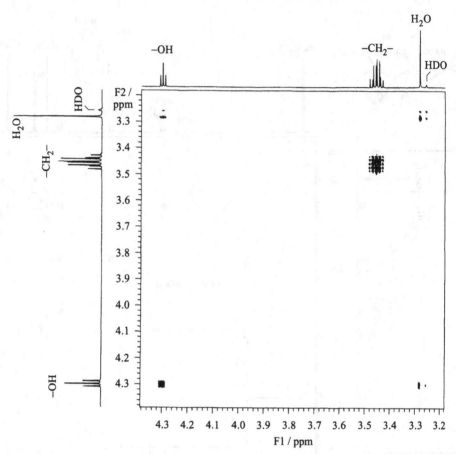

Fig. 6.12. 500 MHz NOESY spectrum of ethanol measured in DMSO-d_6 at 35 °C (the CH$_3$ resonance region is not shown). Recycle time 5 s, 16 scans, spectral width 1,828.7 Hz, 2, 048 data points (*F2*), 512 data points (*F1*), mixing time 5.3 s

All the NOESY cross-peaks mentioned previously have opposite phase to the diagonal peaks, indicating that these arise from positive NOE. NOESY cross-peaks sharing the same phase as the diagonal appear between the protons that exchange chemically with each other. So the NOESY spectrum is effective in studying chemical exchange of the protons. The partial NOESY spectrum of ethanol in DMSO-d_6 at 35 °C is shown in Fig. 6.12. The signals of the hydroxyl protons in C_2H_5OH, H_2O, and HDO appear at 4.29, 3.27, and 3.25 ppm. Three sets of cross-peaks are observed as shown in the figure, indicating the chemical exchange among these protons.

6.6 Conformation Analysis by 2D NMR

Two-dimensional NMR can be applied to the analysis of the conformation of molecules as well as their chemical structures. As described in Sect. 6.2 (Figs. 6.5, 6.6), the cross-peak of the α-CH$_3$ proton with one of the α-CH$_2$ protons at 1.90 ppm is stronger in intensity than that with the α-CH$_2$ protons 1.426 ppm, indicating that one of the α-CH$_2$ protons (H$_A$ or H$_B$ in Structure 6.3) assumes the W conformation to the α-CH$_3$ proton. Then, the following two conformations are possible about the C$_1$–C$_2$ backbone bond as displayed by Newman projections (Structure 6.4).

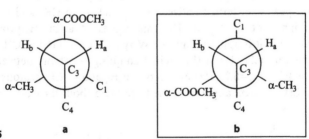

Structure 6.3   MMA *racemo* dimer

Structure 6.4

Structure 6.5

The NOESY spectrum (Fig. 6.11) shows that t-C$_4$H$_9$ protons have connectivities with α-CH$_3$, α-CH$_2$ (H$_A$ and H$_B$), and α-OCH$_3$ protons, respectively. In the conformation depicted in Structure 6.4a, no cross-peaks are expected between t-C$_4$H$_9$ and α-OCH$_3$ protons. The real conformation must be that of Structure 6.4b, and the α-CH$_2$ signals at 1.426 and 1.90 ppm can be assigned to H$_A$ and H$_B$ protons, respectively.

The next problem is the conformation about the C_2-C_3 bond. As described in Sect. 6.2 and Fig. 6.6, one of the ω-CH$_2$ signals at 2.39 ppm shows connectivity with the α-CH$_3$ signal at 1.31 ppm but the other ω-CH$_2$ signal at 1.434 ppm does not. This means that the possible conformations about the C_2-C_3 bond are as shown in Structure 6.5.

As described in Sect. 6.2, the C_4-ω-CH$_3$ bond adopts a gauche conformation to both the C–H bonds of the ω-CH$_2$ group as depicted by Structure 6.6. Then, the coupling constant between H_b and ω-CH protons that are trans to each other should be larger than that between H_a and ω-CH protons that are gauche to each other. The coupling constant between the ω-CH proton and one of the ω-CH$_2$ protons that shows an NMR signal at 2.39 ppm is 8.27 Hz, while the coupling constant between the ω-CHproton and the other ω-CH$_2$ proton at 1.434 ppm is 3.49 Hz (Fig. 6.13) [9, 10]. The results indicate that the ω-CH$_2$ signal at 2.39 ppm can be assigned to H_b and that at 1.434 ppm to H_a.

Structure 6.6

The α-CH$_3$ proton adopts the W conformation to the H_a proton in the conformation shown in Structure 6.5a and to H_b in the conformation shown in Structure 6.5b. The ^{1}H COSY connectivity between the α-CH$_3$ proton and the ω-CH$_2$ protons is very much stronger with the H_b proton than with the H_a proton (Figs. 6.5, 6.6). This clearly indicates that the conformation about the C_2-C_3 bond is that of Structure 6.5b. The long-range enhanced ^{1}H COSY spectrum measured in nitrobenzene-d_5 at 90 °C (Fig. 6.14b [10], Sect. 6.2, Fig. 6.6) clearly shows that a stronger cross-peak is observed between the signals of ω-CH$_3$ and H_a than that between the signals of ω-CH$_3$ and H_b. This suggests that the H_a proton adopts the conformation that is approaching the W-type one with the ω-CH$_3$ proton as shown in Structure 6.7b. From the vicinal coupling constants between H_b and ω-CH protons ($^3J_{Hb-\omega\text{-}H}$=8.27 Hz) and between H_a and ω-CH protons ($^3J_{Ha-\omega\text{-}H}$= 3.49 Hz), the displacement angle from the trans state about the C_3-C_4 bond could

Structure 6.7   **a** (30 °C)   **b** (90 °C)   **c** (110 °C)

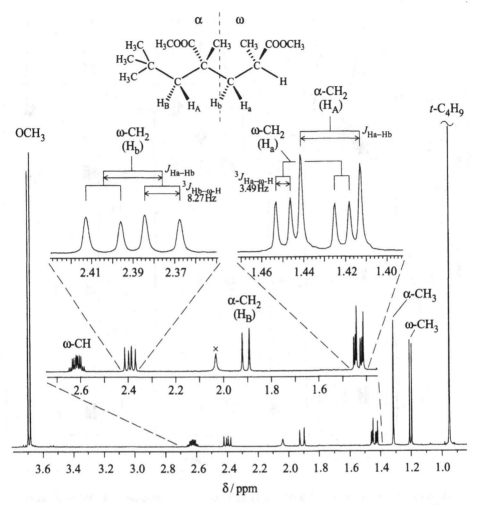

Fig. 6.13. 500 MHz ^{1}H NMR spectrum of the racemo dimer of MMA measured in nitrobenzene-d_5 at 90 °C [9, 10]. 45° pulse, pulse repetition time 10 s, 32 scans

be calculated by using the Karplus correlation [11] to be 18° and 12°, respectively, i.e., the displacement angle may be around 15° anticlockwise.[3] When the COSY spectrum was measured at 110 °C, the cross-peak between ω-CH$_3$ and H$_a$ became much stronger, suggesting that the displacement angle is increased compared with the conformation of Structure 6.7b (Structure 6.7c). The values of $^3J_{Hb-\omega-H}$ and $^3J_{Ha-\omega-H}$ are 8.15 and 3.74 Hz, respectively. The Karplus correlation [11] indicates the displacement angle to be 20–13° anticlockwise.

---

[3] Coupling constants between protons on vicinal carbon atoms depend on the dihedral angle (φ) between the H–C–C and C–C–H planes and the correlation has been graphed by Karplus [11].

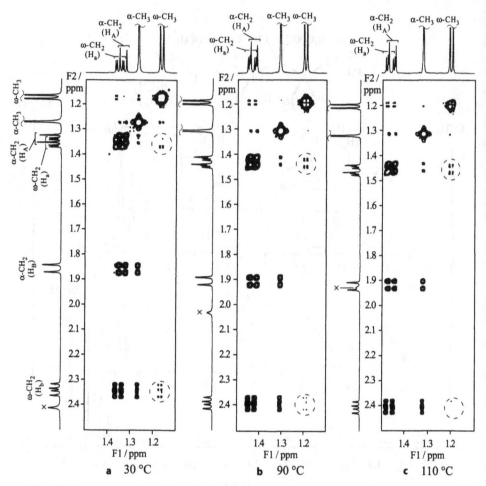

Fig. 6.14. Long-range enhanced 500 MHz ^{1}H COSY spectra of the racemo dimer of MMA measured in nitrobenzene-d_5 **a** at 30 °C, **b** at 90 °C, and **c** at 110 °C [10]. Recycle time 6 s, 32 scans, spectral width 1,623.7 Hz, 2,048 data points (*F2*), 256 data points (*F1*). A delay time of 0.2 s was adopted for long-range enhancement

When measured at 30 °C, the COSY spectrum (Fig. 6.14a) shows that the cross-peak between ω-CH$_3$ and H$_a$ is weaker than that between ω-CH$_3$ and H$_b$. This suggests that the displacement angle changes from anticlockwise to clockwise. The displacement angle calculated from the coupling constants ($^3J_{Hb-\omega-H}$=9.07 Hz and $^3J_{Ha-\omega-H}$=3.06 Hz) is from 7° to 9° anticlockwise. The results are roughly consistent with the COSY spectrum data (Fig. 6.14a) if the errors of the data estimated from the Karplus relation are taken into account. The conformational energy calculation for the dimer was executed to confirm the NMR results [12].

In the conformer analysis, the observed vicinal couplings are often assumed as an average over the conformations present [13]. The existence of the cross-peaks between H_a and ω-CH_3 and between H_b and ω-CH_3 suggests the contribution of the conformers (Structure 6.8b, c) in addition to Structure 6.8a, which in principle is identical to Structure 6.7a. H_b and ω-CH_3 may be in the W form in the conformer in Structure 6.8c, while H_a and ω-CH_3 may be in the W form in the conformer in Structure 6.8b. The increase of the cross-peak between H_b and ω-CH_3 at 30 °C (Fig. 6.14a) suggests the enhanced contribution of the conformer in Structure 6.8c at lower temperature, in which H_a and ω-H are in a trans conformation and H_b and ω-H are in a gauche conformation. If the population of this conformer increases at low temperature, the $^3J(H_a-\omega$-H) value should increase and the $^3J(H_b-\omega$-H) value should decrease. The observed changes are not consistent with these expectations, however. The results thus imply that the dominant population of the conformer in Structure 6.8a is enhanced at lower temperature. However, the different temperature dependence of the two minor conformers (Structure 6.8b, c) might affect the temperature-dependent change in the average conformations shown in Structure 6.7.

Structure 6.8

The conformational analysis for the meso dimer has been made similarly by 2D NMR spectroscopy and the results are shown in Structure 6.9 [10]. The ω-CH_2 protons (H_a and H_b) in the meso dimer are both flanked by a carbonyl group only on one side (Structure 6.9b, c). On the other hand, one of the ω-CH_2 protons, H_b, in the racemo dimer is flanked by carbonyl groups on both sides, while the H_a proton is flanked by no carbonyl groups (Structures 6.5b, 6.6); i.e., the magnetic environments of these two protons are largely different. This is the reason why the chemical shift difference between H_a and H_b in the racemo dimer (0.96 ppm) is larger than that in the meso dimer (0.29 ppm). The chemical shift differences

Structure 6.9

between the two α-CH_2 protons (H_A and H_B) are very similar in the meso and racemo dimers, the reason for which is very understandable from Structures 6.4b and 6.9a.

Thus, 2D NMR spectroscopy is very useful for the structure and conformation analyses of oligomers and polymers.

6.7   Incredible Natural Abundance Double Quantum Transfer Experiment

The 2D INADEQUATE provides direct carbon–carbon connectivities and enables us to sketch the carbon skeletons of organic compounds with the least ambiguity. The technique utilizes directly attached carbon–carbon coupling ($J_{^{13}C-^{13}C}$) with the aid of double-quantum filtering. The probability that any two adjacent carbon atoms are both ^{13}C atoms is about 10^{-4}. This requires rather high sample concentrations even with modern high-field NMR spectrometers, and more than 50 wt/vol% solution is usually needed to obtain a good spectrum. The extremely low sensitivity of 2D INADEQUATE is the only problem of this powerful technique.

The 125 MHz INADEQUATE spectrum of (E)-isophorone diisocyanate in $CDCl_3$ at 35 °C at a concentration of 50 wt/vol% is shown in Figure 6.15 [9]. In the figure, the abscissa (F2 axis) is the carbon axis and is given in ppm for the ^{13}C NMR chemical shifts. The ordinate (F1 axis) corresponds to the frequencies of double quantum transfer. It is usually given in hertz and twice the range of the F2 axis.[4] A set of two cross-peaks appears between the two directly bonded carbons. The cross-peaks are actually doublets with a spacing equal to the coupling constant between the two carbons and are connected horizontally with each other as observed in the figure. The collection of the midpoints for all the lines that connect the two sets of cross-peaks lies on a line running along the diagonal. This can be used to distinguish genuine cross-peaks from spurious peaks and other artifacts.

Peak assignments for the ^{13}C NMR spectrum of (E)-isophorone diisocyanate can be made from Fig. 6.15 in the following way. A quaternary carbon should show four cross-peaks, and accordingly the peaks at 36.70 and 31.74 ppm are both attributed to quaternary carbons, C5 and C3. Since C5 has the electron-withdrawing group ($-CH_2NCO$), the peak appearing at lower magnetic field (36.7 ppm) is assignable to C5, and thus the peak at 31.74 ppm is assignable to C3. The four cross-peaks of C5 connect horizontally with the peaks at 50.85, 45.97, 43.86, and 29.82 ppm. The cross-peaks at the lowest field (50.85 ppm) have only one cross-peak, namely the reciprocal connection to C5, and then the peak at 50.85 ppm can be assigned to the methylene carbon of $-CH_2NCO$ on C5. The peak at 29.82 ppm can be assigned similarly to the methyl carbon on C5 (5-CH_3). The cross-peak at 45.97 ppm has connectivity to the quaternary carbon C3 and the peak can be assigned to the C4 methylene carbon connecting C3 and C5. Consequently the peak

---

[4] For further explanation, see Ref. [2]).

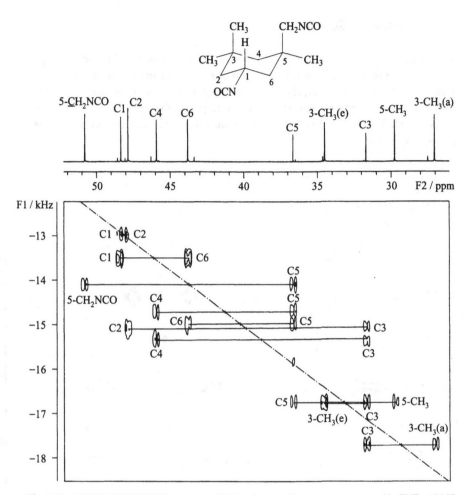

Fig. 6.15. 125 MHz INADEQUATE spectrum of (E)-isophorone diisocyanate, measured in CDCl₃ at 35 °C at a concentration of 50 wt/vol% [9]. Recycle time 7 s, 64 scans. F1 spectral width 13,300 Hz, 2,048 data points. F2 spectral width 26,600 Hz, 128 data points. 1/(4 J_{C-C}) 4.6 ms

at 43.86 ppm is ascribed to the C6 methylene carbon. Then the peaks at 48.40 and 47.90 ppm can be assigned to C1 and C2, respectively, by tracing one carbon's connectivities in due order (C6→C1→C2→C3). Thus all six carbons, C1–C6, in the cyclohexane ring are connected as shown in the figure. The peak assignments for the two methyl carbons on the C3 quaternary carbons were made from the general rule that the axial carbon resonates at higher field than the equatorial one, and is consistent with the result of the NOESY spectrum, where the C1 proton shows NOE correlation with the C3-axial CH₃ protons [14]. Each cross peak between the C1 and C2 carbons does not appear as the doublet reflecting the spin-coupling constant. This is due to the fact that the chemical shifts of these two carbons are very close to each other.

6.8 *J*-Resolved Spectroscopy

J-resolved spectroscopy can clearly display the peak multiplicities due to spin–spin coupling by presenting chemical shifts on one axis and the multiplicities on the other axis. In the spectrum, the splitting by spin-coupling is presented at right angles to the conventional spectrum and overlapping absorptions can sometimes be resolved. The method can be divided into homonuclear and heteronuclear *J*-resolved spectroscopies.

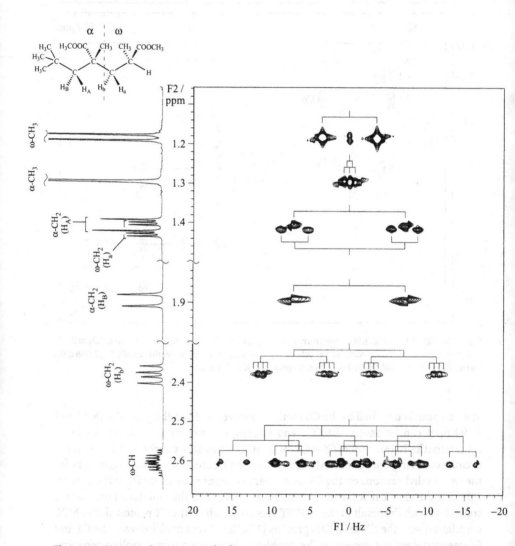

Fig. 6.16. 500 MHz homonuclear ^{1}H–^{1}H *J*-resolved spectrum of the racemo dimer of MMA measured in nitrobenzene-d_5 at 90 °C [10]. Recycle time 10 s, 32 scans. *F2* spectral width 1,371.2 Hz, 2,048 data points. *F1* spectral width 50 Hz, 256 data points

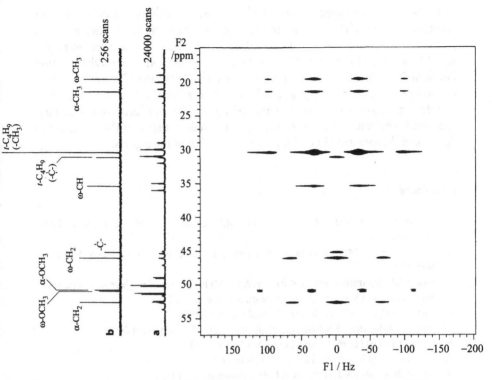

Fig. 6.17. 125 MHz heteronuclear ^1H–^{13}C J-resolved spectrum of the racemo dimer of MMA measured in nitrobenzene-d_5 at 90 °C [10]: **a** nondecoupled and **b** decoupled ^{13}C NMR spectra. Recycle time 5 s, 128 scans. F2 spectral width 5,502.4 Hz, 1,024 data points. F1 spectral width 500.0 Hz, 64 data points

The J-resolved ^1H–^1H spectrum of the racemo MMA dimer measured in nitrobenzene-d_5 at 90 °C is shown in Fig. 6.16 [10]. The multiplet signal of ω-CH at 2.62 ppm is really resolved into 16 peaks as illustrated in the figure, indicating that the splittings result from the spin coupling by the three ω-CH_3 protons and two nonequivalent ω-CH_2 protons (4×2×2). The signal of ω-CH_2 (H_b) at 2.39 ppm is resolved into the quartet peaks, each of which again splits into a small quartet. These small splittings arise from the long-range coupling with α-CH_3 protons as described in Sects. 6.2 and 6.6 (Fig. 6.6), and cannot be observed in the conventional NMR spectrum. Although the α-CH_3 signal at 1.31 ppm appears as a singlet in the one-dimensional spectrum, the splittings due to the long-range coupling are clearly observed in the J-resolved spectrum. The existence of the long-range spin-coupling of α-CH_3 protons with CH_2 protons H_b and H_B was evidenced from the ^1H COSY spectrum (Fig. 6.5). The multiplet signal at about 1.4 ppm is the overlap of ω-CH_2 (H_a, quartet) and α-CH_2 (H_A, doublet) and this can be seen clearly in the resolved spectrum. Thus, the J-resolved ^1H–^1H spectrum provides detailed information on spin–spin coupling between protons and also the exact values of the coupling constants.

In the ^1H–^{13}C heteronuclear J-resolved spectrum, ^{13}C chemical shifts are presented on one axis and ^1H–^{13}C spin-coupling on the other. The ^1H–^{13}C heteronuclear J-resolved spectrum of the racemo MMA dimer measured in nitrobenzene-d_5 at 90 °C is shown in Fig. 6.17 [10]. In the figure decoupled (Fig. 6.17b) and non-decoupled (Fig. 6.17a) ^{13}C NMR spectra are shown. Multiplicity due to ^1H–^{13}C spin-coupling for each carbon signal can be clearly seen in the J-resolved spectrum, and thus the type of carbon (CH_3, CH_2, CH, and quaternary carbons) can be easily identified. The result of the spectrum is equivalent to that available from a non-decoupled ^{13}C spectrum but without severe overlap of the signals.

References

1. CLARIDGE TDW (1999) High-resolution NMR techniques in organic chemistry. Pergamon, New York
2. DEROME AE (1987) Modern NMR techniques for chemistry research. Pergamon, New York
3. ERNST RR, BODENHAUSEN G, WOKAUN A (1987) Principles of nuclear magnetic resonance in one and two dimensions. Oxford Science, Oxford
4. CHENG HN, LEE GH (1990) Trends Anal Chem 9:285
5. MIRAU PA, SHARON SA, KOEGLER G, BOVEY FA (1991) Polym Int 26:29
6. WERSTLER DD (1996) Compr Polym Sci 2nd Suppl 197
7. TONELLI AE (1997) Annu Rep NMR Spectrosc 34:185
8. UTE K, NISHIMURA T, HATADA K (1989) Polymer J 21:1027
9. HATADA K (1992) Kaigai Koubunshi Kenkyu 124
10. KAWAUCHI T, TERAWAKI Y, KITAYAMA T, HATADA K (to be published)
11. KARPLUS M (1959) J Chem Phys 30:11
12. MAEDA K, YASHIMA E, KAWAUCHI T, TERAWAKI Y, KITAYAMA T, HATADA K (to be published)
13. BOVEY FA (1969) Polymer conformation and configuration. Academic, New York, chap 3
14. HATADA K, UTE K (1987) J Polym Sci Part C Polym Lett 25:477

7 NMR Relaxation

In NMR measurements, the nuclei that become excited by the irradiating RF field give up their surplus energy to their surroundings over a period of time to return to the equilibrium state. This process is called NMR relaxation, and involves longitudinal relaxation and transverse relaxation. The time constants of these relaxation processes, known as the spin–lattice relaxation time (T_1) and the spin–spin relaxation time (T_2), are important NMR parameters besides chemical shifts and spin-coupling constants. In this chapter the basic principles of NMR relaxation and its practical application to polymers are described. A detailed description of NMR relaxation and its application for general purposes are found in Refs. [1–7].

7.1 Basic Principles of NMR Relaxation

The magnetization of the nuclei by a pulse begins to return to its original equilibrium value along the z-axis and in the xy-plane immediately after the pulse (Fig. 7.1). The return of the z-component (M_z) to its equilibrium value (M_0) is longitudinal relaxation and the return of M_{xy} to zero is transverse relaxation. Both processes are first-order processes and can be characterized by the times T_1 and T_2, with Eqs. (7.1) and (7.2), respectively:

$$dM_z/dt = -(M_z - M_0)/T_1, \tag{7.1}$$

$$dM_x/dt = -M_x/T_2, \, dM_y/dt = -M_y/T_2. \tag{7.2}$$

In T_2 relaxation, no energy is transferred from the nuclei to their surroundings and the process is an entropy-driven one. Such an entropy-driven process is the exchange of spins between neighboring nuclei and thus the T_2 relaxation is often called spin–spin relaxation. T_1 relaxation, however, always involves loss of energy from the nuclei to the surroundings and is called spin–lattice relaxation. Because M_{xy} can return to zero before M_z regains its equilibrium value but M_z can never return to equilibrium before M_{xy} becomes zero, the value of T_2 is not longer than T_1, i.e., $T_2 \leq T_1$. For many neat liquids and solutions in which molecular motion is sufficiently rapid, T_1 is equal to T_2 and the measurement of T_1 can be made accurately and easily compared with that of T_2, which requires more elaborate pulse techniques [1–7]. Therefore, only spin–lattice relaxation is described in this chapter.

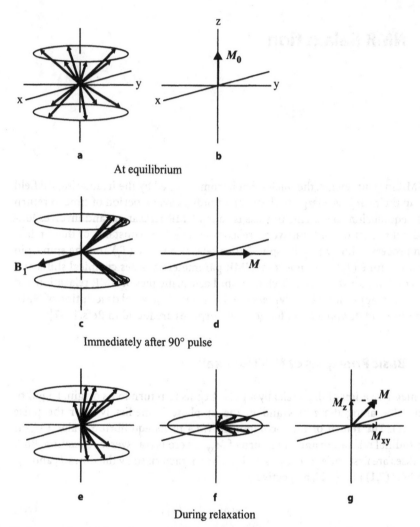

Fig. 7.1. a–g Precessing nuclei at equilibrium and during relaxation. **a** Nuclei precessing at equilibrium, **b** magnetization at equilibrium, **c** precessing nuclei immediately after 90° pulse, **d** magnetization immediately after 90° pulse, **e** nuclei losing coherence and relaxing toward a Boltzmann distribution, **f** components in the *xy*-plane fanning out as relaxation occurs, **g** magnetization and components during relaxation [3]

In order for the T_1 relaxation to occur, it is necessary for a torque to be exerted on the nuclei. This torque should be provided by a magnetic field rotating at the Larmor frequency of the nuclei, which is in the range 10^7–10^9 Hz. All kinds of molecular motions (translation of molecules, rotation of molecules, internal rotation of groups, etc.) which involve the movement of charged particles, electrons, and nuclei exert the torque necessary to allow the nuclei to relax back to their more stable state. The extent to which the nuclei can interact with magnetic fields depends on the magnitude and frequencies of the fields. Rotational motions are usually characterized by their correlation time, τ_c, which is defined as the average time for rotation by 1 rad for rotational motion. The values of τ_c depend on the size and shape of the molecule, temperature, and the viscosity of the solution, and are 10^{-13}–10^{-11} s for small molecules and may increase to about 10^{-9} s for macromolecules in solution. A short event whose duration is t (seconds) has frequencies distributed around $1/t$ (Hertz). Therefore, the rotational motion of a macromolecule with $\tau_c=10^{-10}$ s has a range of frequencies around 10^{10} Hz. Some of these frequencies are at the Larmor frequency and permit interaction with a magnetic field rotating at the Larmor frequency.

There are a variety of reorientational processes that produce magnetic fields and the experimentally measured relaxation rate, $1/T_{1exp}$, is the sum of the rate for all the possible relaxation mechanisms as shown in Eq. (7.3), where the subscripts DD, SR, SA, SC, and E represent the dipole–dipole, spin rotation, shielding anisotropy, scalar coupling, and electron (paramagnetic) relaxation mechanisms, respectively:

$$1/T_{1exp} = 1/T_{1DD} + 1/T_{1SR} + 1/T_{1SA} + 1/T_{1SC} + 1/T_{1E}. \qquad (7.3)$$

Among the possible mechanisms the important two, dipole–dipole and spin–rotation mechanisms, will be described here. The dipole–dipole mechanism is the most important for nuclei with $I=1/2$ such as ^1H and ^{13}C. The magnetic field at a nucleus from a neighboring particle (either a nucleus or an electron with a magnetic moment) depends on the magnetic moment of the particle and the distance between them. The interaction of the rotating field due to a magnetic dipole is greater for particles with larger magnetic moments. For example, an unpaired electron has a magnetic moment almost 1,000 times larger than that of a proton and is much more efficient in producing spin–lattice relaxation than a proton. Thus, the unpaired electron (paramagnetism) of molecular oxygen dissolved in a sample is one of the major sources of relaxation, particularly in the case of small molecules. On the other hand, the proton has a much larger gyromagnetic ratio than the ^{13}C nucleus and is more efficient than other nuclei for the relaxation of ^{13}C in organic compounds.

The local magnetic field induced by the magnetic dipole of neighboring nuclei falls off rapidly with increasing distance between the interacting nuclei. The nuclei that are directly attached are much more effective for relaxation. The probability that a ^{13}C nucleus attaches directly to another ^{13}C nucleus in an organic compound is extremely low owing to the low natural abundance of ^{13}C (1.1%). As a result, the proton attached to the carbon in organic molecules effectively induces spin–lattice relaxation of the ^{13}C nucleus. The relaxation rate for ^{13}C depends on the number of attached protons; CH_2 carbons relax more rapidly than CH carbons, which in turn relax more rapidly than quaternary carbons. T_1 for a CH_3 carbon is usually longer than the expected value ($T_{1CH_3}=T_{1CH}/3$). This can be explained by considering the dynamics of the CH_3 group; the CH_3 group has greater freedom of motion and rotates more rapidly and thus has a shorter correlation time. The shorter correlation time results in a rotating magnetic field with higher frequency which is apart from the Larmor frequency, and in less efficient relaxation. Thus, the dipolar relaxation time, T_{1DD}, gives important information on the chain dynamics, such as relative motion of molecular segments.

The second most important process for spin–lattice relaxation is the spin-rotation mechanism, which is due to the fact that molecules consist of many charged particles (nuclei and electrons) that produce a rotating magnetic field when the molecule rotates. Generally, the relaxation process becomes more efficient as the rotation becomes faster. Hence, the spin-rotation mechanism is most important for small molecules and for quaternary carbon nuclei in which the dipole–dipole mechanism is inefficient.

The measurement of the spin–lattice relaxation time is preferably followed by the measurement of the NOE since the effect is intimately connected with the dipolar relaxation by Eq. (7.4):

$$NOE = 1 + \frac{\gamma_H T_1}{2\gamma_C T_{1DD}} = \frac{T_1}{T_{1DD}}. \tag{7.4}$$

NOE is observed as an increase in the intensities of ^{13}C resonances with proton decoupling, and is defined as the ratio of the intensities with and without the decoupling. From a historical point of view, NOE first made possible the rapid development of ^{13}C NMR spectroscopy.

Equation (7.4) shows that the maximum NOE value of 2.99 is obtained when the relaxation is purely dipolar, i.e., proton decoupling increases the intensities of the ^{13}C resonances by a factor of 2.99. Under this condition, the longer T_1 of carbon certainly indicates the faster motion of the molecular segment when the number, n, of protons attached directly to the carbon is the same. If the value of n is different, the value of nT_1 should be compared instead of T_1 for the mobility. If relaxation mechanisms other than dipole–dipole interaction are involved, T_1 is smaller than T_{1DD}, and the enhancement factor is reduced. The contribution of the dipolar relaxation, T_{1DD}, can be easily determined from the observed value of T_1 using Eq. (7.4). Under decoupling conditions and for purely dipolar relaxation in the extremely narrowing limit ($\omega_0^2 \tau_C^2 \ll 1$), T_{1DD} of carbon can be related to τ_C of the carbon nucleus by Eq. (7.5):

$$\tau_C = \gamma_C^{-2} \gamma_H^{-2} h^{-2} / (nT_{1DD} r^{-6}). \tag{7.5}$$

where n is the number of hydrogen atoms which interact with the carbon nucleus, and r is the internuclear distance between carbon and hydrogen.

7.2 Measurement of T_1 and NOE

As mentioned in Sect. 2.1, the values of the relaxation times must be taken into consideration in establishing the parameters in NMR measurements, particularly pulse width and pulse repetition. T_1 of a particular nucleus can also give information on the structural details, such as the relative motion of molecular segments, steric hindrance, and the proximity of groups. Therefore, accurate and reliable measurement of T_1 is one of the important tasks in NMR experiments.

The most widely applied method to determine T_1 is the inversion-recovery method [8], on which all the other methods [9-15] are ultimately based. In this procedure the magnetization vector of a spin system initially in equilibrium is inverted into the negative z-direction by a 180° pulse. Following a delay time t, during which the magnetization undergoes relaxation to a new value, the vector is turned by a 90° pulse into the xy-plane, and the FID is then recorded. The pulse sequence can be written as $(-180°-t-90°-T-)_n$. The delay time, T, should be at least 5 times the longest T_1 to be determined in the experiment. The experiment is repeated for a series of different values of t. Measurements with a t value of $5T_1$ should be included at least twice, at the beginning and at the end of each experiment to obtain accurate results. The set of data consisting of peak intensities is then processed by the integrated form of Eq. (7.1) (Eq. 7.6):

$$\ln (M_z - M_0) = -t/T_1 + C. \tag{7.6}$$

From the boundary condition that, $M_z = -M_0$ at $t = 0$, the value of the constant C is calculated to be $\ln(-2M_0)$ and Eq. (7.6) can be rewritten as Eq. (7.7):

$$\ln (M_z - M_0) = -t/T_1 + \ln (-2M_0). \tag{7.7}$$

Equation (7.7) can be rewritten as Eq. (7.8):

$$\ln (M_0 - M_z) = -t/T_1 + \ln (2M_0). \tag{7.8}$$

In practice the intensity of each peak of interest is then measured for each different time, t, and is assumed to be proportional to M_z. Then, Eq. (7.8) can be rewritten again as Eq. (7.9):

$$\ln (I_\infty - I_t) = -t/T_1 + \ln (2I_\infty). \tag{7.9}$$

where I_t is the peak intensity at time t and I_∞ is the peak intensity when equilibrium has been reestablished (as measured with $t \geq 5T_1$, $I_0 = I_\infty$). T_1 is the reciprocal of the slope of the linear plot of $\ln(I_\infty - I_t)$ against t. A typical set of experiments is shown in Fig. 7.2a for radically polymerized PMMA in toluene-d_8 at 100 °C. From the plots of $\ln(I_\infty - I_t)$ against t, T_1 values of CH_2, α-CH_3, OCH_3, C-4, and C=O carbons were determined as 0.18, 0.27, 1.39, 3.00, and 1.54 s, respectively, as shown in Fig. 7.2b.

A rough estimate of T_1 can be made from the t for the null value of I_t (t_n) using Eq. (7.10) (see the circled points in Fig. 7.2a):

$$T_1 = t_n/\ln 2. \tag{7.10}$$

The estimated T_1 can be used to set an appropriate value of T in the pulse sequence $(-180°-t-90°-T-)_n$.

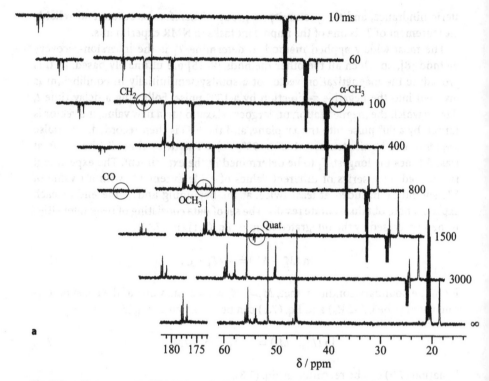

Fig. 7.2. a Measurement of ^{13}C T_1 for radically prepared PMMA in toluene-d_8 (1.0 mol/l) at 100 °C and 125 MHz. Set of ^{13}C NMR spectra measured by inversion recovery (Fig. 7.2b *see p. 175*)

NOE for the carbons in a molecule can be determined from wide-band proton-decoupled ^{13}C spectra. First, a spectrum is obtained with continuous wide-band proton decoupling. A second spectrum is obtained by using pulse-modulated wide-band proton decoupling, in which the decoupler is gated on only during data acquisition periods. In the latter experiment, the ^{13}C spectrum is obtained without NOE. In these measurements the pulse repetition time should be longer than 5 times T_1 for the carbon with the longest T_1. The value of the NOE can be calculated by dividing each individual peak integration in the continuously decoupled spectrum by the peak integration in the spectrum measured without NOE. A typical example of NOE determination is shown in Fig. 7.3 for radically prepared PMMA in toluene-d_8 at 100 °C and 125 MHz.

The NOE values are 2.18 for CH_2, 2.77 for α-CH_3, 2.11 for OCH_3, 1.88 for quaternary carbon (Quat C), and 1.20 for C=O. Except for α-CH_3, the values are less than the theoretical maximum (2.99). The NOE value for C=O reflects some contribution of chemical shift anisotropy [2]. Although the NOE values for CH_2, OCH_3, and Quat C are determined mainly by a dipolar mechanism, relatively slow motion of the polymer segment, i.e., a longer correlation time (τ_c), contributes to the decrease of NOE when the resonance frequency (ω) becomes large. When the same

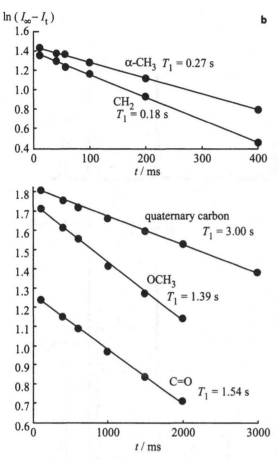

Fig. 7.2. b Measurement of ^{13}C T_1 for radically prepared PMMA in toluene-d_8 at 100 °C and 125 MHz. $\ln(I_0 - I_t)$ against t. Delay time 10 s, 800 scans

sample was measured at 25 MHz, the NOE values for all the carbons were close to 2.99, and evidently indicate that the spin–lattice relaxation mostly arises from the dipolar mechanism. The ^{13}C NMR resonance at 25 MHz for the polymer molecule should be in the extremely narrowing limit ($\omega^2\tau^2 \ll 1$) since the values of τ for polymer molecules are in the range 10^{-9}–10^{-11} [16–18]. Then, the T_1 values mentioned previously (Fig. 7.2b) indicate that the mobility of the carbon increased in the order $CH_2 < \alpha\text{-}CH_3 < OCH_3$ since the nT_1 values are in the same order. The higher mobility of the α-CH_3 group than the main-chain CH_2 group is due to the internal rotation of the CH_3 group which reduces the effectiveness of dipolar coupling to directly bonded protons. The much higher mobility of the OCH_3 group results from the greater freedom of the internal rotation [19].

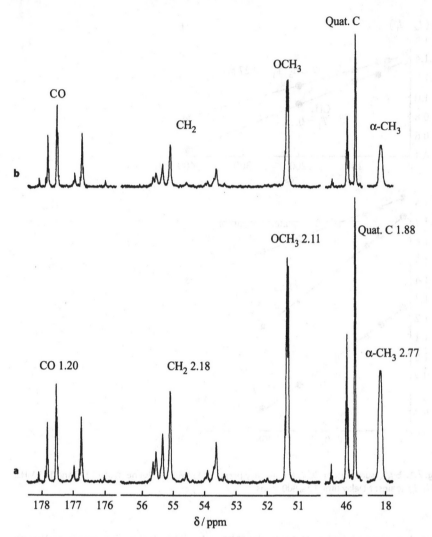

Fig. 7.3a,b. Determination of NOE factors for ¹³C NMR signals of radically prepared PMMA in toluene-d_8 (1.0 mol/l) at 100 °C and 125 MHz. The *numbers* indicate the NOE values of the carbons. **a** with NOE, **b** without NOE

7.3   Sample Preparation – Effect of Oxygen and Type of Sample Tube

In order to make a reliable measurement of T_1, the sample solution must be carefully degassed, particularly in the case of low-molecular-weight compounds (large T_1), since dissolved oxygen is paramagnetic and has a great effect on the relaxation time. Removal of oxygen can be accomplished by several freeze–pump–thaw cycles under vacuum followed by sealing the solution in the tube under vacuum or dry

Table 7.1. ^{1}H T_1 and ^{13}C T_1 of methyl isobutyrate in toluene-d_8 at 55 °C at 500 and 125 MHz, respectively, under air, nitrogen, and a vacuum

	^{1}H T_1 (s)			^{13}C T_1 (s)		
	Air	N$_2$	Vacuum	Air	N$_2$	Vacuum
CO	–	–	–	31.63	39.47	40.77
OCH$_3$	5.66	17.37	17.07	14.96	19.80	18.51
CH	7.45	44.06	42.53	26.86	31.09	30.74
CH$_2$	4.81	11.36	11.01	11.16	12.06	12.28

nitrogen or argon. NMR tubes have very thin walls and sometimes are broken during the freeze–thaw procedure by the use of a liquid nitrogen bath. To avoid such a risk, the use of a dry ice/acetone bath is recommended for degassing followed by warming to ambient temperature in order to evacuate dissolved gas from the solution to the gas phase. The procedure should be repeated several times. As to sealing the tube under nitrogen or argon, see Sect. 1.4.3. Dissolved oxygen can also be removed by bubbling dry nitrogen or argon into the solution.

The ^{1}H T_1 and ^{13}C T_1 of methyl isobutyrate measured in toluene-d_8 at 55 °C and 500 MHz under air, nitrogen, and a vacuum are given in Table 7.1. The T_1 measured in nitrogen $[T_1(N_2)]$ is very similar to that measured under vacuum $[T_1(vac)]$, but is much longer than the T_1 measured in air $[T_1(air)]$. The relaxation contribution from dissolved oxygen $[T_1(O_2)]$ can be estimated as 8–9 s for ^{1}H T_1 and as 100–200 s for ^{13}C T_1 using Eq. (7.11). If we allow ±5% error for T_1 determination, degassing is necessary only for samples where T_1 is longer than 0.5 s in ^{1}H relaxation and longer than 5 s in ^{13}C relaxation. Actually, the measurements of ^{1}H T_1 and ^{13}C T_1 for PMMA under air give almost the same T_1 as those measured under vacuum or nitrogen (Table 7.2).

$$1/T_1(air) = 1/T_1(vac) + 1/T_1(O_2).$$ (7.11)

However, a polymer sometimes degrades at high temperatures in air depending on the structure of the polymer and solvent; therefore, the measurement under nitrogen is recommended. The measurements under vacuum at high temperature sometimes cause refluxing of the solvent, which results in the instability of the measurement. This point should be carefully considered when you wish to make measurements under vacuum at high temperatures. For example, the determination of ^{13}C T_1 in toluene-d_8 (boiling point 110 °C/760 mmHg) under vacuum at 105 °C fluctuated owing to instability in the course of the measurement.

In the case of long T_1, there is some possibility that some of the spins excited by the 180° pulse move out from the effective part of sample volume before being magnetized by the 90° pulse, resulting in the incorrect determination of T_1. Short T_1 may be observed compared with the T_1 expected for the measurement without the

Table 7.2. ^{1}H T_1 and ^{13}C T_1 of isotactic PMMA in toluene-d_8 at 500 and 125 MHz, respectively, under air, nitrogen, and a vacuum

	^{1}H T_1 (s) at 100 °C				^{13}C T_1 (s) at 90 °C		
	Air	N$_2$	Vacuum		Air	N$_2$	Vacuum
OCH$_3$	1.427	1.533	1.539	OCH$_3$	1.849	1.887	1.815
				CO	2.636	2.713	2.605
CH$_2$ (2.33 ppm)	0.393	0.388	0.396				
CH$_2$ (2.30 ppm)	0.386	0.391	0.391				
				CH$_2$	0.284	0.280	0.281
CH$_2$ (1.61 ppm)	0.348	0.356	0.353				
CH$_2$ (1.58 ppm)	0.345	0.349	0.351				
				Quat C	4.130	4.048	4.051
α-CH$_3$	0.443	0.450	0.455	α-CH$_3$	0.469	0.463	0.458

diffusion. Use of a small-volume sample tube such as that shown in Fig. 1.18d may prevent this kind of inaccurate measurement. As shown in Table 7.3, the ^{13}C T_1 values of PMMA measured in toluene-d_8 at 55 and 80 °C are independent of the type of sample tube. However, in the measurements at 100 °C the ^{13}C T_1 measured in a small-volume tube are, particularly those for carbonyl and quaternary carbon, a little but meaningfully longer than that measured in a standard tube, indicating that the measurements are affected by the diffusion. The ^{1}H T_1 that are shorter than the ^{13}C T_1, showed very similar values in the measurements in both types of sample tube even at 100 °C.

7.4 Reliability of Spin–Lattice Relaxation Time and NOE

As mentioned previously, relaxation parameters are essential in adjusting data acquisition conditions in FT-NMR measurement to obtain quantitative data. In particular, reliable determinations of T_1 and NOE are required for the correct setting of measurement conditions.

Among the numerous matters that require attention for T_1 measurement, the most important is the accurate adjustment of 180° and 90° pulse lengths. Use of inaccurate pulse lengths results in shorter T_1.

T_1 values depend on the frequency of measurement, temperature, solvent, and solution concentration and these should be clearly and accurately quoted with the data. Details about these matters are described in Sect. 7.5.

It is also important to know whether or not the accuracy of the determination of the relaxation parameters allows us to treat them as one of the characteristics of a polymer. From this point of view, ^{1}H T_1 and ^{13}C T_1 values and ^{13}C NOE values of radically prepared PMMA($\bar{M}_n$=28,500, $\bar{M}_w/\bar{M}_n$=2.12, $mm:mr:rr$=4.0:34.7:61.3) in

Table 7.3. Determination of ^1H T_1 and ^{13}C T_1 for isotactic PMMA in toluene-d_8 under nitrogen at various temperatures at 500 and 125 MHz using standard (std) and small-volume (small) NMR tubes

^{13}C-T_1 (s)

	55 °C		80 °C		100 °C	
	Std	Small	Std	Small	Std	Small
OCH$_3$	1.30	1.22	1.51	1.51	1.86	1.90
CO	1.91	1.87	2.51	2.41	3.79	4.22
CH$_2$	0.21	0.20	0.26	0.25	0.36	0.37
Quat C	3.19	3.03	3.68	3.66	4.23	5.31
α-CH$_3$	0.29	0.29	0.40	0.40	0.58	0.58

^1H-T_1 (s)

	80 °C		100 °C	
	Std	Small	Std	Small
OCH$_3$	1.30	1.29	1.47	1.44
CH$_2$ (2.33 ppm)	0.35	0.33	0.39	0.39
CH$_2$ (2.30 ppm)	0.35	0.35	0.39	0.38
CH$_2$ (1.61 ppm)	0.31	0.31	0.35	0.34
CH$_2$ (1.58 ppm)	0.31	0.31	0.35	0.33
α-CH$_3$	0.38	0.37	0.45	0.45
t-C$_4$H$_9$	1.12	1.10	1.48	1.43

Table 7.4. ^{1}H T_1's (ms) of PMMA measured in $CDCl_3$ at 55 °C at different frequencies [20]. The figures in parentheses represent the standard deviation σ (%)

Frequency (MHz)	n^a	α-CH$_3$		CH$_2$	OCH$_3$
		rr	mr	rrr	
60	1	50 (–)	60 (–)	40 (–)	140 (–)
90	3	72 (4.0)	84 (4.9)	65 (3.3)	307 (6.9)
100	3	74 (1.3)	87 (1.1)	70 (2.7)	382 (9.0)
100^b	2	58 (4.3)	66 (4.5)	67 (7.5)	354 (1.1)
200	5	113 (3.9)	132 (5.4)	138 (1.0)	560 (7.5)
200^b	2	104 (0)	108 (1.8)	168 (0.3)	662 (0.3)
270	2	142 (0.1)	162 (7.4)	200 (4.7)	851 (–)
360	1	188 (–)	205 (–)	275 (–)	963 (–)
400	7	208 (1.3)	220 (2.7)	318 (0.9)	1133 (9.8)
500	4	259 (2.3)	260 (2.5)	389 (6.9)	1166 (20.5)

[a] Number of determinations.
[b] Measurements were made at room temperature (24 and 27 °C).

$CDCl_3$ were collected from 27 NMR spectrometers by the Research Group on NMR Society of Polymer Science, Japan, through a round robin method [20]. The observing frequencies ranged from 60 to 500 MHz for ^{1}H NMR and from 15 to 125 MHz for ^{13}C NMR. The mean values and standard deviations of the ^{1}H T_1 of PMMA determined in $CDCl_3$ at 55 °C at different resonance frequencies are summarized in Table 7.4. The standard deviations for the ^{1}H T_1 values of α-methyl and backbone methylene protons are mostly less than 5%. These T_1 values can be regarded as characteristic data for the polymer as long as they are obtained under the specified conditions. The ^{1}H T_1 data for methoxy protons showed fairly large standard deviations, the reason for which is not clear at present. The ^{1}H T_1 values of all sorts of protons increase with increasing resonance frequency as discussed in Sect. 7.5.

The mean values and standard deviations for the ^{13}C T_1 of the PMMA measured in $CDCl_3$ at 55 °C at different frequencies are listed in Table 7.5. The standard deviations for ^{13}C T_1 are larger than those for ^{1}H T_1 and exceed 10% in some cases. Scattering of the data does not seem to be noticeably different with regard to the types of carbon, though the standard deviation for OCH$_3$ is slightly larger than for others, similarly to the case of ^{1}H T_1 for OCH$_3$.

^{13}C NOE is also an important relaxation parameter and directly affects the signal intensities in the ^{13}C NMR spectrum. The assessment of NOE data and its precision is sometimes desired for quantitative analysis. The mean values and standard deviations for the NOE of PMMA determined in $CDCl_3$ at 55 °C at different frequencies are summarized in Table 7.6. The precision of the NOE measurements is as good as that of ^{13}C T_1. The NOE values obtained at or below 25 MHz

Table 7.5. ^{13}C T_1 (ms) of PMMA measured in $CDCl_3$ at 55 °C at different frequencies [20]. The figures in parentheses represent the standard deviation σ (%)

Frequency (MHz)	n^a	α-CH₃		Quat C		OCH₃	CH₂	C=O		
		rr	mr	rr	mr		rrr	mmrr+rmrr	rrrr	rrrm
15	1	80	100	700	460	460	40	1600	1720	840
22.5	4	74(10.6)	111(10.6)	813(5.0)	938(3.8)	23(6.9)	56(4.8)	2273(3.6)	1938(2.5)	2050(5.5)
25	4	71(9.3)	94(12.0)	907(12.6)	912(6.9)	604(6.5)	54(7.7)	2303(11.7)	1951(11.6)	2068(7.5)
50	5	87(7.2)	120(6.6)	1136(10.3)	1238(12.9)	661(19.7)	79(4.1)	2274(17.8)	1864(8.6)	1880(9.2)
50^b	2	61(7.4)	88(5.7)	1038(1.7)	1242(16.3)	567(6.2)	72(0.7)	1803(2.6)	1663(1.0)	1632(4.8)
67.5	1	115	145	1409	1544	861	105	2001	1740	1713
90.6	1	123	165	1820	1790	793	118	1450	1450	2200
100	6	129(8.5)	161(9.5)	1870(2.6)	1988(4.0)	973(9.7)	139(3.2)	1453(6.0)	1323(5.2)	1406(7.6)
125	4	134(5.2)	165(6.3)	2237(4.6)	2362(3.7)	1157(16.3)	156(6.7)	1168(5.7)	1075(6.9)	1164(7.9)

^{a} Number of determinations.
^{b} Measurements were made at room temperature (24 °C).

Table 7.6. ^{13}C NOE of PMMA measured in CDCl$_3$ at 55 °C at different frequencies [20]. The figures in parentheses represent the standard deviation σ (%)

Frequency (MHz)	n^a	α-CH$_3$		Quat C		OCH$_3$	CH$_2$	C=O		
		rr	mr	rr	mr	mr	rrr	mmrr+rmrr	rrrr	rrrm
15	1	2.80	2.40	2.60	2.10	2.70	2.80	2.60	2.70	2.22
22.5	3	2.64(10.2)	2.63(14.3)	2.57(4.5)	2.36(5.9)	2.39(4.8)	2.35(13.2)	2.10(10.5)	2.19(4.5)	1.83(5.4)
25	4	2.72(1.6)	2.70(4.5)	2.47(1.1)	2.55(4.4)	2.52(4.1)	2.72(9.2)	2.30(9.8)	2.24(5.8)	2.21(5.5)
25^b	1	2.75	2.75	1.95	1.95	1.89	2.09	1.62	1.62	1.62
50	5	2.65(4.1)	2.65(5.1)	2.11(7.6)	2.17(3.7)	2.12(5.1)	2.13(10.9)	1.73(9.8)	1.62(6.6)	1.61(11.7)
50^b	2	2.21(1.2)	2.36(1.2)	1.75(4.4)	1.86(1.9)	1.80(3.1)	1.80(3.1)	1.43(1.5)	1.39(2.0)	1.39(1.0)
67.5	2	2.31(6.3)	2.42(2.6)	1.88(8.3)	1.84(6.2)	1.88(7.2)	1.83(7.7)	1.35(10.2)	1.29(6.6)	1.23(3.5)
90.6	1	2.07	2.03	1.98	1.97	1.66	1.71	1.59	1.33	1.33
100	5	2.14(10.0)	2.20(6.9)	1.68(4.3)	1.75(5.6)	1.63(4.9)	1.63(6.0)	1.09(6.6)	1.07(5.9)	1.10(8.1)
125	4	2.16(4.0)	2.30(7.7)	1.68(6.2)	1.56(11.7)	1.57(7.5)	1.63(8.5)	1.11(10.4)	1.07(3.8)	1.12(9.9)

^{a} Number of determinations.
^{b} Measurements were made at room temperature (24 °C).

Table 7.7. [1]H T_1 (500 MHz) [13]C T_1 (125 MHz), and ^{13}C NOE (125 MHz) of isotactic PMMA measured in toluene-d_8 at 55 °C by a single person using a single spectrometer [26]. The figures in parentheses represent the standard deviation σ (%)

Signal	T_1 (s)[a]		NOE[a]
	[1]H	[13]C	
CO		1.71 (2.2)	1.13 (3.1)
CH$_2$ (2.33 ppm)	0.38 (0.2)		
CH$_2$ (1.57 ppm)	0.31 (0.2)	0.20 (1.4)	2.12 (1.9)
OCH$_3$	1.40 (0.9)	1.14 (1.7)	1.88 (2.7)
Quat C		3.01 (2.5)	1.79 (3.0)
α-CH$_3$	0.31 (0.3)	0.27 (2.1)	2.71 (1.8)

[a] Mean values for five determinations.

are about 2.5–2.8, which are close to their theoretically expected maxima for all but carbonyl carbon, and decrease with increasing resonance frequency as seen in Table 7.6 and discussed in Sect. 7.5.

When the measurements of T_1 and NOE were made by a single person with a single spectrometer, scattering of the results became much less than that of the results obtained by different persons with different spectrometers. The results obtained for isotactic PMMA by Yoshio Terawaki using a JEOL GX500 NMR spectrometer (500 MHz) installed at Osaka University, Faculty of Engineering Science, Toyonaka, Osaka, Japan, are given in Table 7.7. The standard deviations for the [1]H T_1, the [13]C T_1 and NOE of isotactic PMMA measured in toluene-d_8 at 55 °C are within 0.9, 2.5, and 3.1%, respectively.

7.5 Effect of Frequency, Temperature, Solution Concentration and Solvent on the Measurement of T_1 and NOE

[1]H T_1 and [13]C T_1 of polymers usually show a strong frequency dependence. The [1]H T_1 of PMMA are plotted in Fig. 7.4 [20] against the observing frequency using data from Table 7.4. The [1]H T_1 values linearly increase with an increase in observing frequency for each proton but the slope of the straight line is considerably smaller for the α-CH$_3$ protons compared with the others. As a result, the α-CH$_3$ protons exhibit shorter T_1 than CH$_2$ protons at frequencies higher than 100 MHz, although the CH$_3$ protons have more nearest-neighbor protons than CH$_2$ protons. However, it is very complicated to analyze these frequency dependences in terms of simple models for the proton magnetic relaxation, because cross-relaxation is significant among these protons [19, 21].

Higher static magnetic fields are preferable for obtaining spectra with higher S/N ratios. However, Fig. 7.4 suggests that the increase in [1]H T_1 at higher magnetic

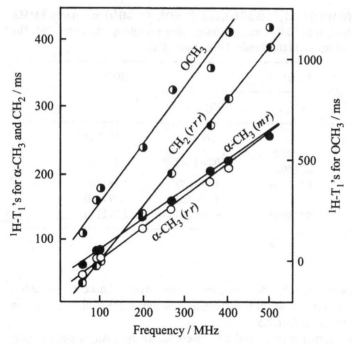

Fig. 7.4. ^{1}H T_1 of radically prepared PMMA measured in CDCl$_3$ at 55 °C at different frequencies [20]

field strength requires setting a longer repetition time for the measurements. This would result in a decrease in the practical S/N ratio obtainable within a limited period of time (see Sect. 2.1).

The resonance frequency dependences of the mean values of ^{13}C T_1 (Table 7.5) are illustrated in Fig. 7.5a and b. ^{13}C T_1 for all carbons except for carbonyl carbon increase linearly with increasing resonance frequency. The α-CH$_3$ carbon in the rr triad shows longer T_1 than the CH$_2$ carbon in the rrr tetrad at lower frequencies but shorter T_1 at higher frequencies, and the inversion in the order of magnitudes of ^{13}C T_1 for α-CH$_3$ and CH$_2$ carbons occurs at a frequency between 67.5 and 90.6 MHz.

^{13}C T_1 for the carbonyl carbon decrease with increasing resonance frequencies above 25 MHz. The relaxation induced by chemical shift anisotropy is possible for the carbon in the π bond. The relaxation by this mechanism becomes important at higher magnetic field, since the contribution of chemical shift anisotropy to the relaxation rate is proportional to the square of the static magnetic field strength if the molecular tumbling rate is much larger than the resonance frequency [22]. The relaxation of the acetylenic carbons in 1,4-diphenylbutadiyne is a typical example of this case [23]. The great decrease in ^{13}C T_1 of the carbonyl carbon in PMMA with increasing magnetic field indicates an increasing contribution of the chemical shift anisotropy to the spin–lattice relaxation [20].

Longer T_1 in ^{13}C NMR at higher resonance frequencies may result in the great disadvantage of requiring a longer repetition time between rf pulses, resulting in

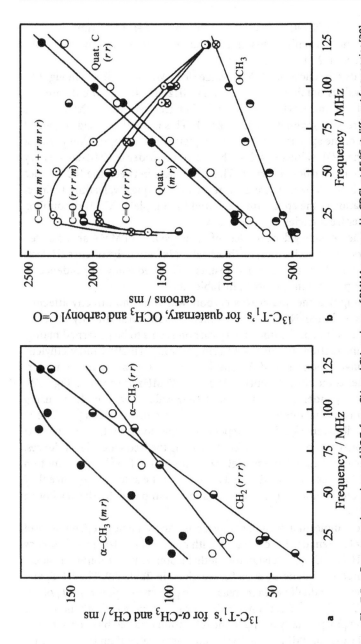

Fig. 7.5. a Frequency dependence of ^{13}C T_1 for α-CH_3 and CH_2 carbons of PMMA measured in $CDCl_3$ at 55 °C at different frequencies [20]. **b** Frequency dependence of ^{13}C T_1 for quaternary, OCH_3, and carbonyl carbons of PMMA measured in $CDCl_3$ at 55 °C at different frequencies [20]

a decreased practical S/N ratio obtainable in a limited period of time, because ^{13}C NMR measurement usually needs more iterative accumulation than ^{1}H NMR measurement (see Sect. 2.1).

The frequency dependences of ^{13}C NOE values for PMMA are shown in Fig. 7.6a and b. The NOE values obtained at or below 25 MHz are close to the theoretical maximum for all but carbonyl carbons. At higher frequencies the NOE values decrease with increasing magnetic field strength. This is another disadvantage of the measurement of the spectrum at higher magnetic field strength. Compared at each frequency, the NOE value of each carbon is in the order α-CH$_3$>quaternary carbon $\cong$ OCH$_3$ $\cong$ CH$_2$>carbonyl carbon. The differences between the NOE values of these carbons increase with increasing resonance frequency, leading to inaccurate intensity ratios in the spectrum measured by complete ^{13}C-^{1}H decoupling with NOE at higher frequencies.

The trends of the frequency dependences of T_1 and NOE for PMMA are observed for other polymers, such as other polymethacrylates [24] and polystyrenes (Hatada K, Kitayama T, Terawaki Y unpublished results). The frequency dependence of ^{13}C T_1 for isotactic polystyrene is shown in Table 7.8.

It is worth noting that the quaternary carbon, which has no directly attached hydrogens, shows a similar NOE value to other carbons with directly attached hydrogens. The relaxation of the quaternary carbon seems to be governed mainly by dipole–dipole interaction with its neighboring protons in methyl and methylene groups. The carbonyl carbon shows distinctively smaller NOE values than other carbons and the values come close to unity at 100 and 125 MHz as shown in Fig. 7.6b.

The frequency dependences of ^{13}C T_1 and NOE values were analyzed using various molecular motional models; box type, log χ^2, 2τ, and 3τ models [25]. The 3τ model, which describes three kinds of independent superposed motions of the C–H internuclear vector, seems most suitable for describing the frequency dependences. However, multiple correlation time models with more than four kinds of motion are considered to be necessary for detailed analyses of the side groups since they should be subjected to additional inner rotations compared to the backbone carbons [25].

In the case of low-molecular-weight compounds, the frequency dependences of T_1 and NOE are not so remarkable compared with macromolecules. The frequency dependences of ^{1}H T_1, ^{13}C T_1, and NOE for methyl isobutyrate, a monomer model of PMMA, measured in toluene-d_8 at 55 °C are given in Table 7.9. The ^{1}H T_1 and ^{13}C T_1 values increase gradually with increasing frequency except for the T_1 of the carbonyl carbon. The T_1 for the carbonyl carbon does not increase with increasing magnetic field strength, probably owing to the relaxation induced by chemical shift anisotropy as in the case of PMMA. The NOE values are smaller than the theoretical maximum (2.99), and show frequency dependence, suggesting the contributions of relaxation mechanisms other than dipole–dipole interaction, such as spin rotation and chemical shift anisotropy.

T_1 values are expected to increase with increasing temperature of measurement since segmental mobility of macromolecules is usually enhanced with temperature. In fact, the ^{13}C T_1 values of PMMAs in toluene-d_8 increase with an increase in tem-

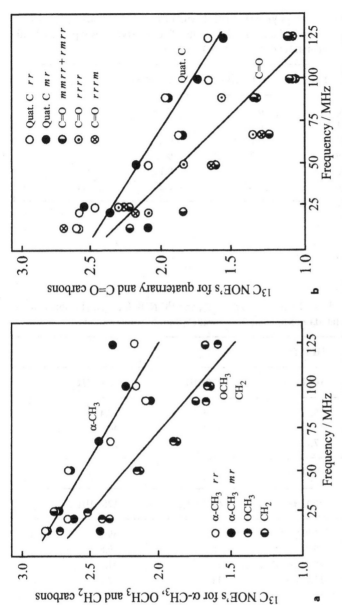

Fig. 7.6. a Frequency dependence of ^{13}C NOE for α-CH$_3$, OCH$_3$, and CH$_2$ carbons of PMMA measured in CDCl$_3$ at 55 °C at different frequencies [20]. **b** Frequency dependence of ^{13}C NOE for quaternary and carbonyl carbons of PMMA measured in CDCl$_3$ at 55 °C at different frequencies [20]

Table 7.8. ^{13}C T_1 of isotactic polystyrene in $CDCl_3$ (3 wt/vol%) at 45 °C at various frequences [Hatada K, Kitayama T, Terawaki Y unpublished results]. The polystyrene was prepared with $(C_2H_5)_3Al$–β–$TiCl_4$ in n-heptane at 60 °C. $\bar{M}_n$ (GPC) = 68,900

Carbon[a]	67.8 MHz	100 MHz	125 MHz
C_1	1.08	0.72	0.58
C_2	0.21	0.24	0.26
C_3	0.20	0.24	0.25
C_4	0.19	0.22	0.23
CH_2	0.12	0.13	0.14
CH	0.18	0.22	0.22

Table 7.9. Frequency dependence of ^1H T_1, ^{13}C T_1, and ^{13}C NOE for methyl isobutyrate in toluene-d_8 at 55 °C [Hatada K, Kitayama T, Terawaki Y unpublished results]

Hydrogen	^1H T_1 (s)		
	100 MHz	270 MHz	500 MHz
OCH_3	9.1	10.5	17.4
CH	13.5	16.4	44.1
CH_3	7.2	8.3	11.4

Carbon	^{13}C T_1 (s)		
	25 MHz	67.5 MHz	125 MHz
CO	41.8	42.2	39.5
OCH_3	15.0	18.2	19.8
CH	25.4	28.9	31.1
CH_3	11.7	11.7	12.1

Carbon	NOE		
	25 MHz	67.5 MHz	125 MHz
CO	1.175	1.067	1.022
OCH_3	1.971	2.171	2.156
CH	1.880	2.467	2.520
CH_3	2.462	2.710	2.946

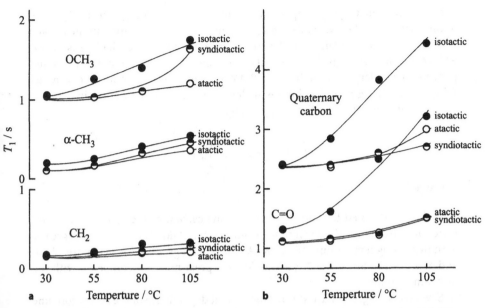

Fig. 7.7a,b. ^{13}C T_1 of isotactic, syndiotactic, and atactic PMMA measured in toluene-d_8 at 125 MHz at different temperatures (Hatada K, Kitayama T, Terawaki Y unpublished results)

perature as shown in Fig. 7.7. The temperature dependence is affected strongly by the tacticity of the polymer, particularly in the case of the T_1 of quaternary and carbonyl carbons which are more greatly increased with increasing temperature in the isotactic polymer than in the syndiotactic one (Fig. 7.7b). The relaxation for quaternary and carbonyl carbons, which have no directly bonded protons, is contributed by the neighboring protons and the contribution is affected by the mobility and conformation of the segment.

The ^1H T_1 of the PMMAs measured in toluene-d_8 increased with increasing temperature, similarly to the ^{13}C T_1 (Hatada K, Kitayama T, Terawaki Y unpublished results). However, ^1H T_1 of α-CH$_3$ and CH$_2$ protons of isotactic and syndiotactic PMMAs measured in nitrobenzene-d_5 showed a different temperature dependence (Fig. 7.8). The T_1 values of the α-CH$_3$ protons in the isotactic PMMA increase with increasing temperature, while the T_1 values of the main-chain CH$_2$ protons in the syndiotactic PMMA decreased. In the Bloembergen–Purcell–Pound theory, T_1 shows a minimum along with the correlation time (τ_c); with decreasing τ_c (i.e., increasing mobility), T_1 decreases in the slow-motion region and increases in the fast-motion region. The phenomenon of the increasing T_1 of the α-CH$_3$ protons in the isotactic PMMA with increasing temperature means that the α-CH$_3$ group has relatively high mobility (small correlation time) and the T_1 value lies in the fast-motion region. The α-CH$_3$ group in the syndiotactic PMMA shows a minimum value at 70 °C, which is a transition point from the slow-motion region to the fast-motion one. At the minimum T_1, τ_c is close to the reciprocal resonance frequency divided by 2π (5.9×10^{-10} s) [2]. The ^1H T_1 of the CH$_2$ group in syndiotactic PMMA decreases with increasing temperature up to 110 °C, indicating that the T_1 minimum exists above 110 °C, while the T_1 minimum for the CH_2 group in the isotactic PMMA is at 70 °C. These results are consistent with the fact that the isotactic PMMA has higher segmental mobility than the syndiotactic PMMA. The temperature dependences of the ^1H T_1 of PMMAs in toluene-d_8 and nitrobenzene-d_5 suggest that the segmental mobilities are lower in nitrobenzene-d_5 than in toluene-d_8, particularly for the syndiotactic polymer.

$$t\text{-}C_4H_9\!-\!\!\left(\!-CH_2\!-\!\underset{\underset{\displaystyle OCH_3}{\overset{\displaystyle C=O}{|}}}{\overset{\overset{\displaystyle CH_3}{|}}{C}}\!-\!\right)_{\!n}\!\!-H$$

Structure 7.1

It should be noted that the T_1 of the methyl carbon of the t-butyl end group introduced from the initiator into the syndiotactic PMMA chain increases greatly with increasing temperature (Fig. 7.8). As the t-butyl group is located at the chain end as shown in Structure 7.1, it should have great mobility compared with the in-chain carbons.

Several investigators have studied the solvent dependence of the relaxation time [24, 26–29] and some of them have discussed the results in terms of solubility parameters. The ^{13}C T_1 of the side-chain carbons of poly(n-BuMA) decrease with increasing difference between the solubility parameters of the polymer and the

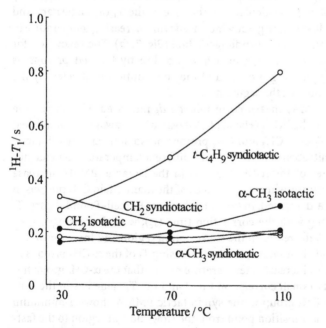

Fig. 7.8. 270 MHz ^1H T_1 of isotactic and syndiotactic PMMAs in nitrobenzene-d_5 at different temperatures (Hatada K, Kitayama T, Terawaki Y unpublished results)

Table 7.10. ^{13}C T_1's of isotactic and syndiotactic PMMAs measured in toluene-d_8 at 105 °C and 125 MHz at different concentrations [Hatada K, Kitayama T, Terawaki Y unpublished results]

Carbon	Isotactic (s)			Syndiotactic (s)		
	5%	10%	20%	5%	10%	20%
CO	3.33	3.27	2.90	1.61	1.43	1.36
CH$_2$	0.31	0.31	0.28	0.19	0.19	0.18
OCH$_3$	1.77	1.74	1.70	1.18	1.13	1.14
Quat C	4.49	4.36	3.69	2.95	2.76	2.67
α-CH$_3$	0.52	0.53	0.49	0.25	0.24	0.24

solvent, $\Delta\delta$, indicating that intermolecular interactions between the polymer molecules are more important in the side-chain relaxation in poor solvents [24]. The ^{13}C T_1 of polystyrene are longer in THF than in dioxane, which corresponds to the fact that $\Delta\delta$ for THF (0.8) is smaller than $\Delta\delta$ for dioxane (1.0) [27]. However, some contradictory results were reported on the relation between the solubility parameter difference and the ^{13}C T_1 of polymers [29]. The solvent dependence of T_1 is still a complex problem.

T_1 values are also affected by the concentration of the solution and usually decreased with increasing solution concentration. ^{13}C T_1 of isotactic PMMA measured in toluene-d_8 at 105 °C at concentrations of 5 and 10% are similar to each other, but those at 20% decreased slightly probably owing to the reduced segmental mobility and increased intermolecular interaction of PMMA molecules (Table 7.10). The ^{13}C T_1 for syndiotactic PMMA also decreased with increasing concentration, although the pattern of the concentration dependence is slightly different from that for isotactic PMMA (Table 7.10)

7.6 Tacticity Dependence of ^{1}H T_1 and ^{13}C T_1 Values and NOE

NMR relaxation data of stereochemically different nuclei should give valuable information about local molecular motion of stereosequences or stereoregular polymers. Earlier reports on the ^{13}C T_1 values of polymers showed equal relaxation times for the corresponding carbons in different steric configurations for solutions of poly-acrylonitrile [30, 31], polystyrene [30], and poly(vinyl chloride) [31]. The tacticity dependence of the T_1 values of vinyl polymers was first reported by Hatada et al. [32] using PMMA in toluene-d_8 at 110 °C and 100 MHz. Their findings are that the ^{1}H T_1 values are longer for an isotactic PMMA than for a syndiotactic PMMA and are also longer for an isotactic triad than for a syndiotactic triad. A similar dependence of ^{1}H T_1 values on tacticity was found for poly[(R,S)-α-methylbenzyl methacrylate]s, poly[MMA-alt-(methyl α-phenylacrylate)], and poly(α-methylstyrene) [32]. Later, many papers were published on relaxation times in solutions particularly on the ^{13}C T_1 of polymethacrylates differing in tacticity [19, 21, 25, 33–46].

Table 7.11. ^{13}C T_1 values and correlation times, τ_c, of various poly(alkyl-methacrylate)s in toluene-d_8 at 110 °C [19]

Ester group		T_1 (s)				$\tau_c \times 10^{11}$		$\tau_{int} \times 10^{11}$ (s)
		CH_2	α-CH_3	C=O	C-4	CH_2	α-CH_3	
CH_3	Isotactic	0.28	0.49	7.85	2.96	8.4	3.1	5.0
	Syndiotactic	0.10	0.22	3.03	1.63	22.4	7.1	10.3
C_2H_5	Isotactic	0.19	0.42	3.50	2.38	12.1	3.7	5.3
	Syndiotactic	0.086	0.22	2.95	1.34	27.1	7.1	9.6
i-C_3H_7	Isotactic	0.13	0.35	2.88	2.36	18.1	4.5	5.9
	Syndiotactic	0.077	0.18	2.33	1.01	30.2	8.9	12.6
t-C_4H_9	Isotactic	0.074	0.24	2.60	1.07	31.5	6.6	8.3
	Syndiotactic	0.036	0.13	1.84	0.67	64.7	11.8	14.4

The T_1 of carbon in isotactic polymethacrylates have been found to be consistently longer than for the corresponding carbons in syndiotactic polymers [19, 21, 29, 33–39, 45, 46]. The ^{13}C T_1 values for various poly(alkyl methacrylate)s measured in toluene-d_8 at 110 °C and at 25 MHz are shown in Table 7.11 [19]. The NOE for each carbon is close to the theoretical maximum under these conditions, indicating that the relaxation is purely dipolar. Thus, the extremely narrowing condition should be satisfied at 25 MHz. Then, the correlation time, τ_C, can be calculated by Eq. (7.5) as mentioned in the introductory part of this chapter (see Sect. 7.1). The calculated τ_C are also shown in Table 7.11. The results strongly indicate that the isotactic polymer chain has greater segmental mobility than the syndiotactic chain. The segmental mobility seems to decrease with an increase in the bulkiness of the ester group in both the isotactic and syndiotactic polymers. The α-CH_3 group has freedom of internal rotation, whose correlation time, τ_{int}, can be estimated as follows:

$$1/\tau_{int} = 1/\tau_{CH_3} - 1/\tau_{main}, \qquad (7.12)$$

where τ_{CH_3} is the total correlation time for reorientation of the methyl group and τ_{main} is the correlation time for the reorientation of the main chain. We can reasonably assume $\tau_{main} = \tau_{CH_2}$. The calculated values of τ_{int} show that the α-CH_3 group in the isotactic polymer also has greater freedom of internal rotation than that in the syndiotactic one (Table 7.11) [19].

The longer T_1 for isotactic PMMA are observed over a wide temperature range at 125 MHz as shown in Fig. 7.7. The difference of the T_1 for quaternary and carbonyl carbons between isotactic and syndiotactic PMMA increases with increasing temperature. The longer T_1 of isotactic PMMA are also observed over a wide range of observing frequency. The results for isotactic, syndiotactic, and atactic PMMAs are shown in Fig. 7.9.

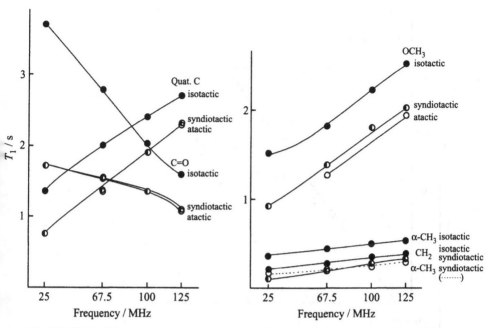

Fig. 7.9. ^{13}C T_1 of isotactic, atactic, and syndiotactic PMMAs in toluene-d_8 at 55 °C at different frequencies (Hatada K, Kitayama T, Terawaki Y unpublished results)

The difference in ^{13}C T_1 between polymers with different tacticities but the same chemical structure results not only from differences in the chain segmental motion between stereoregular polymers or sequences but also from differences in preferred conformations which lead to different average distances for the interaction of a carbon with a proton of a neighbouring monomeric unit [36]. The solvent dependence of ^{13}C T_1 for stereoregular PMMAs has been explained by the solvent-dependent conformation of the polymer chain [37–39].

In the case of ^{1}H T_1, the relaxation mechanism is rather complicated and the observed T_1 cannot be unconditionally related to the segmental mobility of the polymer chains [37–39]. However, the ^{1}H T_1 values of poly(alkyl methacrylate)s measured at 100 MHz were found to be parallel with the corresponding ^{13}C T_1 values measured at 25 MHz for α-CH$_3$ and backbone CH$_2$ groups, i.e., ^{1}H T_1 values can be regarded as a measure of segmental mobility of the polymers. In fact, the ^{1}H T_1 values of the isotactic polymers were always longer than those of the corresponding protons in the syndiotactic polymers (Fig. 7.10) [19].

The tacticity dependence of the ^{13}C T_1 values was also observed for the methine carbons of polypropylene and poly(but-1-ene). Short T_1 values were observed for the syndiotactic polymers compared with the corresponding values for the isotactic polymers. The difference is small but significant [47].

^{13}C T_1 values of isotactic-rich poly(alkyl vinyl ether)s (alkyl=CH$_3$, C$_2$H$_5$, iso-C$_3$H$_7$, iso-C$_4$H$_9$, t-C$_4$H$_9$) were measured in toluene-d_8 at 110 °C and 25 MHz [48]. The T_1 values for syndiotactic sequences are consistently longer than those for

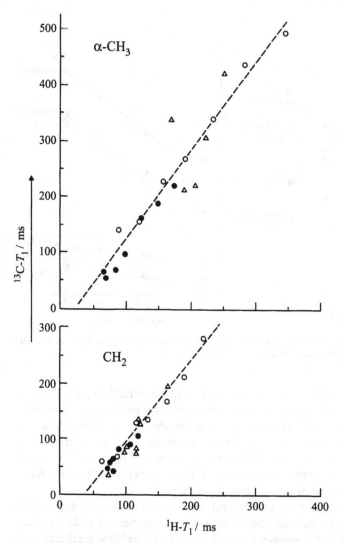

Fig. 7.10. Relationship between ^{13}C and ^{1}H spin–lattice relaxation times, ^{13}C T_1 and ^{1}H T_1, for methyl (*upper*) and methylene (*lower*) groups in isotactic PMMA (*open circles*), syndiotactic PMMA (*closed circles*), and other polymethacrylates (*triangles*) [19]

isotactic sequences, in contrast to the cases of polymethacrylates, polypropylenes, and poly(but-1-ene) mentioned previously. The NOE values for polymers of methyl and t-butyl vinyl ethers are close to the theoretical maximum. These results indicate that the syndiotactic sequences have higher segmental mobility than the isotactic sequences.

The ^{13}C T_1 of 1,4-polyisoprenes and 1,4-polybutadienes were measured in both the solid state and in solution and the tacticity dependences were discussed [49–52].

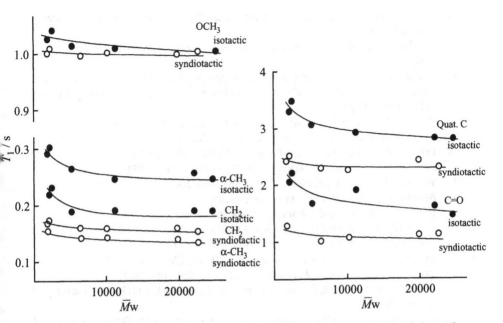

Fig. 7.11. ^{13}C T_1 of isotactic and syndiotactic PMMAs with different molecular weights in toluene-d_8 at 55 °C and 100 MHz

7.7 Molecular Weight Dependence of ^{13}C T_1

As already mentioned the values of ^{13}C T_1 strongly depend on the segmental mobility of the polymer chain and thus depend on the molecular weight of the polymer, particularly in the range of lower molecular weights. The ^{13}C T_1 of isotactic and syndiotactic PMMAs with different $\bar{M}_n$ in CDCl$_3$ at 55 °C and 125 MHz are shown in Fig. 7.11. The T_1 for the isotactic polymer decrease rapidly with increasing molecular weight up to $\bar{M}_n \cong 1.0 \times 10^4$ and are almost constant at higher $\bar{M}_n$. In the case of the syndiotactic polymer, the T_1 are almost constant in the $\bar{M}_n$ range higher than 2,000. The difference in the molecular-weight dependence between the isotactic and syndiotactic polymers may be due to the conformational difference; the isotactic flexible helical chain has greater freedom than the syndiotactic zigzag chain, particularly in the low-molecular-weight region.

References

1. FARRAR TC, BECKER ED (1971) Pulse and Fourier transform NMR: introduction to theory and methods. Academic, New York, p 46
2. LEVY GC (1974) Topics in carbon-13 NMR spectroscopy, vol. 1. Wiley, New York, chap 3
3. YODER CH, SCHAEFFER CP JR (1987) Introduction to multinuclear NMR. Benjamin/ Cummings, Malno Park, CA, p 67

4. DEROME AE (1987) Modern NMR techniques for chemistry research. Pergamon, New York, chaps 4, 5
5. SLICHTER CP (1980) Principles of magnetic resonance. Springer, Berlin Heidelberg, New York
6. ABRAGAM A (1983) Principles of nuclear magnetism. Oxford University Press, Oxford
7. ABRAHAM RJ, LOFTUS P (1978) Proton and carbon-13 NMR spectroscopy. Heyden, Philadelphia, chap 6
8. VOLD RL, WAUGH JS, KLEIN MP, PHELPS DE (1968) J Chem Phys 48:3831
9. LYERLA JR, GRANT DM (1972) Int Rev Sci Phys Chem Ser 1 4:155
10. LYERLA JR, LEVY GC (1974) Top Carbon-13 NMR Spectrosc 1:79
11. BREITMAIER E, SPOHN KH, BERGER S (1975) Angew Chem 87:152
12. WEHRLI FW (1976) Top Carbon-13 NMR Spectrosc 2:343
13. BERGER S (1978) Angew Phys Org Chem 16:239
14. BOERÉ RT, KIDD RG (1982) Annu Rep NMR Spectrosc 13:319
15. CRAIK DJ, LEVY GC (1984) Top Carbon-13 NMR Spectrosc 4:239
16. SCHAEFER J, NATUSCH DFS (1972) Macromolecules 5:416
17. ALLERHAND A, DODDRELL D, KOMOROSKI R (1971) J Chem Phys 55:189
18. DODDRELL D, GLUSHKO V, ALLERHAND A (1972) J Chem Phys 56:3683
19. HATADA K, KITAYAMA T, OKAMOTO Y, OHTA K, UMEMURA Y, YUKI H (1978) Makromol Chem 179:485
20. CHÛJÔ R, HATADA K, KITAMARU R, KITAYAMA T, SATO H, TANAKA Y, HORII F, TERAWAKI Y. Members of research group on NMR, SPSJ (1988) Polym J 20:627
21. HATADA K, ISHIKAWA H, KITAYAMA T, YUKI H (1977) Makromol Chem 178:2753
22. ABRAGAM A (1961) The principles of nuclear magnetism. Oxford University Press, London, chap 8
23. LEVY GC, CARGIOLI JP, AMET FAL (1973) J Am Chem Soc 95:1527
24. LEVY GC, WANG D (1986) Macromolecules 19:1013
25. HORII F, NAKAGAWA M, KITAMARU R, CHÛJÔ R, HATADA K, TANAKA Y, members of research group on NMR, SPSJ (1992) Polym J 24:1155
26. HEATLEY F, WOOD B (1978) Polymer 19:1405
27. GRONSKI W, MURAYAMA N (1978) Makromol Chem 179:1509
28. INOUE Y, KONNO T (1976) Polym J 8:457
29. HATADA K, KITAYAMA T, TERAWAKI Y, OHTA K, OKAMOTO Y, YUKI H, LENZ RW (1988) Bull Inst Chem Res Kyoto Univ 66:311
30. SCHAEFER J, NATUSCH DFS (1972) Macromolecules 5:416
31. INOUE Y, NISHIOKA A, CHÛJÔ R (1973) J Polym Sci Polym Phys Ed 11:2237
32. HATADA K, OKAMOTO Y, OHTA K, YUKI H (1976) J Polym Sci Polym Lett Ed 14:51
33. UTE K, NISHIMURA T, HATADA K (1989) Polym J 21:1027
34. LYERLA JR JR, HORIKAWA TT (1976) J Polym Sci Polym Lett Ed 14:641
35. HATADA K, KITAYAMA T, SAUNDERS K, LENZ RW (1981) Makromol Chem 182:1449
36. LYERLA JR JR, HORIKAWA TT, JOHNSON DE (1977) J Am Chem Soc 99:2463
37. INOUE Y, KONNO T, CHÛJÔ R, NISHIOKA A (1977) Makromol Chem 178:2131
38. INOUE Y, KONNO T (1978) Makromol Chem 179:1311
39. SPEVÁCEK J, SCHNEIDER B (1978) Polymer 19:63
40. HEATLEY F, COX MK (1980) Polymer 21:381
41. HEATLEY F, COX MK (1981) Polymer 22:190
42. HEATLEY F, COX MK (1981) Polymer 22:288

43. ZAJICEK J, PIVCOVÁ H, SCHNEIDER B (1981) Makromol Chem 182:3169
44. ZAJICEK J, PIVCOVÁ H, SCHNEIDER B (1981) Makromol Chem 182:3177
45. ASAKURA T, SUZUKI K, HORIE K, MITA S (1981) Makromol Chem 182:2289
46. OH SH, RYOO R, JHON MS (1989) J Polym Sci Polym Chem Ed 27:1383
47. ASAKURA T, DOI Y (1983) Macromolecules 16:786
48. HATADA K, KITAYAMA T, MATSUO N, YUKI H (1983) Polym J 15:719
49. SCHAEFER J (1972) Macromolecules 5:427
50. KOMOROSKI RA, MAKFIELD J, MANDELKERN L (1977) Macromolecules 10:545
51. GRONSKI W, MURAYAMA N (1976) Makromol Chem 177:3017
52. HATADA K, KITAYAMA T, TERAWAKI Y, TANAKA Y, SATO H (1980) Polym Bull 2:791

8 On-line SEC/NMR Analysis of Polymers

Recent developments in the sensitivity and resolution of NMR spectrometers have permitted us to use them as a real-time detectors for HPLC as demonstrated in recent review articles [1-4]. SEC is the most widely and frequently used HPLC method for polymer analysis. Recently, we developed an on-line SEC/NMR system in which a 500 or 750 MHz ^{1}H NMR spectrometer was used as a detector [5-13]. By using on-line SEC/NMR the molecular weight of a polymer can be determined without a calibration curve if the polymer sample contains a known amount of end groups per polymer molecule. Data on the molecular-weight dependence of polymer structures, such as copolymer composition and tacticity, are very useful for understanding the mechanism of polymerization or the properties of polymers, and are usually collected by the fractionation of the polymer and the subsequent structural analysis of each fraction. However, this method requires a lot of time and rather a large amount of polymer sample. The on-line SEC/NMR method is very useful for this purpose; it needs only a few hours and a very small amount of sample (0.5–1.0 mg).

8.1 SEC/NMR Instrumentation

Our on-line SEC/NMR system consists of a high-performance liquid chromatograph and an NMR spectrometer with a high magnetic field. A 2- or 3-mm inner diameter (i.d.) glass tube with a tapered structure at both ends is employed as an NMR observation flow cell (Fig. 8.1) [6]; the detection volumes are 60 (2-mm i.d.) and 140 µl (3-mm i.d.), respectively. A similar type of NMR flow cell was used in the on-line HPLC/NMR experiments [14, 15]. The tapered structures at both ends are needed to prevent turbulent flow in the cell, which would cause broadening of the elution band. When a cell without tapering was used, some broadening of the band was observed [5]. The flow cell is mounted in the proton probe designed specially for the SEC/NMR system. The connections between the chromatograph equipped with an SEC column and the flow cell are made with 3-m Teflon tubing (0.3-mm i.d.).

Chloroform-d containing 0.5% ethanol-d_6 is usually used as an eluant. The background signals due to the small amounts of impurities in the eluant, except for the H_2O signal, can be eliminated by subtracting the baseline absorbance.

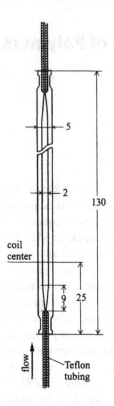

Fig. 8.1. Schematic diagram of the 5-mm (outer diameter) glass tube used for the NMR observation flow cell. The *numbers* indicate lengths in millimeters [6]

Undeuterated solvent containing a small amount of deuterated solvent for deuterium lock can also be used as an eluant. In this case the eluant signals can be eliminated by applying the water suppression enhanced through T_1 effects (WET) [16] in combination with the liquid chromatography NMR probe with pulsed-field-gradient coils [12].

Typical operating conditions of our SEC/NMR experiments are as follows. The ^{1}H NMR data are collected over the entire chromatographic peak and are stored as a consecutive series of n coadded scans. If the repetition time is t s, a series of NMR spectra are obtained every nt s and thus the time resolution of the chromatogram is nt s. The 45–90° pulse is usually used. One of the characteristic features of on-line SEC/NMR is the use of the flow cell. The resolution of the spectrum depends on the flow rate of the sample solution; slight broadening of the resonance is observed at flow rates greater than 0.5 ml/min.

The S/N ratio of the resonance decreases gradually with increasing flow rate. At flow rates greater than 0.5 ml/min the S/N decreases to about one fifth that in the nonflow state. So a flow rate of 0.5 ml/min or below is usually used [10]. It should be noted that the resolution of the spectrum obtained using the flow cell is much better than the resolution obtained in the ordinary measurement without spinning and compares favorably with that in the ordinary measurement with spinning.

The decrease in S/N at flow rates above 0.5 ml/min is mainly due to the insufficient premagnetization of the sample; as the flow rate increases, the residence time of the sample within the NMR cell before signal detection decreases, leading to lower S/N. When a 90° pulse is used, the residence time has to be longer than $5T_{1max}$ for quantitative determination (T_{1max} is the largest T_1 among the T_1 values of protons in the sample). For the signal with $T_1=0.5$ s at the flow rate of 0.5 ml/min the volume of 0.021 ml (=5×0.5×0.5÷60) has to be located in the front of the part of the cell used for signal detection.

8.2 Molecular Weight Determination of Polymers by SEC/NMR

As mentioned in the previous section, determination of molecular weight by SEC requires a calibration curve, which is usually prepared using a set of standard polystyrenes with narrow molecular-weight distribution (MWD). Preparation of other standard polymers with different molecular weights and narrow MWDs is generally difficult or laborious, and much attention has been given to empirical means of deriving calibrations for polymers other than polystyrene [17].

NMR spectroscopy can provide the number-average molecular weight of polymers if the polymer molecule contains a known amount of end groups per chain, and can be used as the detector for SEC that does not require a calibration curve. Here the results of a molecular-weight determination for highly isotactic PMMA by the on-line SEC/NMR technique [10] are shown. The isotactic PMMAs were prepared by the living polymerization of MMA with t-C$_4$H$_9$MgBr in toluene at -78 °C. The polymer molecules contain one t-C$_4$H$_9$ group at the end of the chain (Structure 8.1) [18], and the $\bar{M}_n$ can be determined from the relative intensities of the ^{1}H NMR signals due to t-C$_4$H$_9$ and α-CH$_3$ groups. The isotacticity of the PMMAs ranged from 96.3 to 97.9% in triads.

$$t\text{-C}_4\text{H}_9 \left(\text{CH}_2 - \overset{\displaystyle \text{CH}_3}{\underset{\displaystyle \underset{\displaystyle \text{OCH}_3}{\overset{|}{\text{C=O}}}}{\overset{|}{\underset{|}{\text{C}}}} \right)_n \text{H}$$

Structure 8.1

The SEC/NMR data of the isotactic PMMA with $\bar{M}_n$ of 3,270 determined by ^{1}H NMR are shown in Fig. 8.2. The signals due to $-$OCH$_3$, CH$_2$, α-CH$_3$, and t-C$_4$H$_9$ protons resonate at 3.60, 2.13 and 1.52, 1.20, and 0.86 ppm, respectively. All the cross-sections at the peak positions give ^{1}H NMR-detected SEC chromatograms. The ^{1}H NMR-detected SEC chromatograms from the α-CH$_3$ signals of the four PMMA samples are shown in Fig. 8.3 together with those recorded using a refractive index (RI) detector. The peak profiles of the ^{1}H NMR-detected and RI-detected chromatograms are very similar to each other. The differences in elution times are due to differences in the void volume of the connecting path.

The ^{1}H NMR spectrum of PMMA (sample A) stored as a single file for elution times from 50.1 to 50.4 min (Fig. 8.3) that has S/N ratios large enough to determine $\bar{M}_n$ accurately is illustrated in Figure 8.4. The calculated $\bar{M}_n$ value of this fraction is 3,520. The $\bar{M}_n$ of each fraction can be determined directly from the

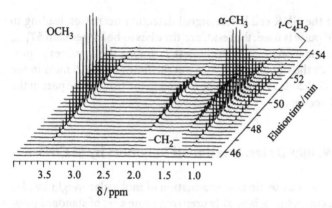

Fig. 8.2. 750 MHz on-line SEC/NMR data of isotactic PMMA. ($\bar{M}_n$ =3,270 from ^{1}H NMR, sample a in Fig. 8.3) [10]. Eluant CDCl$_3$, flow rate 0.2 ml/min, amount of sample 1.0 mg, pulse width 90°, pulse repetition 2.25 s

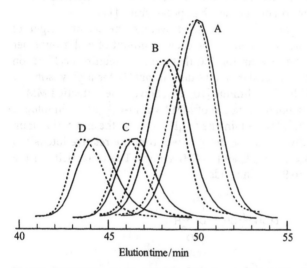

Fig. 8.3. ^{1}H NMR-detected (*solid lines*) and RI-detected (*dotted lines*) SEC curves of isotactic PMMAs. The $\bar{M}_n$ values from ^{1}H NMR of samples A, B, C, and D are 3,270, 5,750, 10,160, and 24,220, respectively. The amounts of samples A, B, C, and D introduced to the chromatograph were 1.0, 0.8, 0.4, and 0.4 mg, respectively, and the volume of the sample solution was 50 ml [10]

^{1}H NMR spectrum.[1] This is one of the great advantages of the on-line SEC/NMR system. Thus the $\bar{M}_n$ and $\bar{M}_w/\bar{M}_n$ values for four PMMA samples were determined from the SEC/NMR data. The results are shown in Table 8.1 together with those obtained by conventional methods. The $\bar{M}_n$ values obtained by SEC/NMR agreed well with those determined from ^{1}H NMR of total polymers and RI-detected

---

[1] The files which exhibit the *t*-C$_4$H$_9$ signal with S/N less than 10 are added with two or three contiguous files so that the ratio exceeds 10.

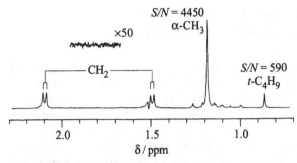

Fig. 8.4. ^{1}H NMR spectrum of isotactic PMMA acquired at the elution maximum of sample A (50.1–50.4 min, see Fig. 8.3) [10]

SEC, particularly for samples A, B, and C. The $\bar{M}_w/\bar{M}_n$ values from SEC/NMR are 5–9% smaller than those obtained by the conventional SEC/RI method. This may be partly due to the small S/N ratios of the signals, particularly the end-group signal at the trailing edges of an elution peak. However, it should be noted that the $\bar{M}_n$ of each fraction determined by SEC/NMR is hardly affected by the instrumental spreading in the SEC since $\bar{M}_n$ is determined without calibration and thus SEC/NMR gives accurate $\bar{M}_w/\bar{M}_n$ values regardless of the spreading [10].

The results mentioned here demonstrate the applicability of 750 MHz SEC/NMR in the analysis of molecular weight and its distribution through quantitative determination of end-group content in a polymer sample. When measurements are carried out under appropriate conditions, reliable data on $\bar{M}_n$ and $\bar{M}_w/\bar{M}_n$ can be obtained up to the average degree of polymerization (DP) of about 200 [10]. When a 500 MHz NMR spectrometer was used, the applicable DP was up to about 100 [6]. So the higher the magnetic field of the spectrometer, the better the results.

Table 8.1. $\bar{M}_n$ and $\bar{M}_w/\bar{M}_n$ of isotactic PMMAs determined by 750 MHz SEC/NMR and conventional methods [10]

Sample	^{1}H NMR[a]	SEC/NMR		SEC/RI[b]	
	$\bar{M}_n$	$\bar{M}_n$	$\bar{M}_w/\bar{M}_n$	$\bar{M}_n$	$\bar{M}_w/\bar{M}_n$
A	3,270	3,350	1.143	2,980	1.231
B	5,750	6,000	1.107	5,400	1.215
C	10,160	10,220	1.088	10,800	1.170
D	24,220	20,540	1.080	28,350	1.135

[a] Determined from the relative intensity of the α-CH$_3$ resonance to the t-C$_4$H$_9$ resonance in the 750 MHz NMR spectrum measured by the use of a 5-mm (outer diameter) probe.

[b] The molecular weight was calibrated against uniform isotactic PMMAs (23-, 40-, and 80-mers) separated by supercritical fluid chromatography.

8.3   Studies on Molecular Weight Dependence of Copolymer Composition and Tacticity

The on-line SEC/NMR technique is also useful for studying the molecular weight dependence of polymer properties such as tacticity [7] and copolymer composition [8, 11, 19]. This type of information is very important for understanding the mechanism of polymerization and is obtained quickly by SEC/NMR compared with the conventional fractionation method. SEC/NMR analysis was carried out for a highly isotactic block copolymer of PMMA and poly(n-BuMA), which has a bimodal MWD. The copolymer was obtained by polymerizing n-BuMA with the PMMA living anion, which was prepared by the polymerization of MMA with t-C_4H_9MgBr in toluene at $-60\,°C$. The stacked trace plot of the serial spectra is shown in Fig. 8.5. The cross-sections at 3.59 ppm (OCH_3 of the MMA unit) and 3.95 ppm (OCH_2 of the n-BuMA unit) give us the changes in the contents of MMA and n-BuMA units, respectively, in the copolymer along with the change in the molecular weight. The results clearly indicate a larger n-BuMA content in the higher-molecular-weight part. Detailed inspection of the results suggested that multiple active species with different activities should be generated in the polymerization of n-BuMA with the PMMA living anions and that the PMMA anions with higher DP have higher reactivity than those with lower DP in the block copolymerization of n-BuMA [8].

SEC/NMR analysis was also made on the copolymers of EMA and t-butyl acrylate prepared with t-C_4H_9Li/bis(2,6-di-t-butylphenoxy)methylaluminum in toluene at $-60\,°C$, and monomer-selective and living character of the copolymerization was evidenced [19].

Ethylene–propylene–(2-ethylidene-5-norbornene) terpolymers (EPDM) (Structure 8.2) were investigated by 750 MHz SEC/NMR using chloroform-d as an eluant (0.2 ml/min). Comparison of the chromatograms obtained by monitoring the CH_2 and CH_3 resonances in the terpolymer sample showed clearly that ethylene-to-propylene ratio depends on the molecular weight and increased with increasing molecular weight. The molecular-weight dependence of small amounts (0.6–1.1 mol%) of 2-ethylidene-5-norbornene units in the terpolymers could not be determined by the typical on-line flow SEC/NMR. However, the stop-and-flow measurements allow larger numbers of accumulations which are enough to determine the norbornene contents and these were found to increase with increasing molecular weight in a typical commercial EPDM having a broad MWD. The E/Z ratio in the norbornene units was constant (72/28) over the whole molecular-weight range [11].

Structure 8.2

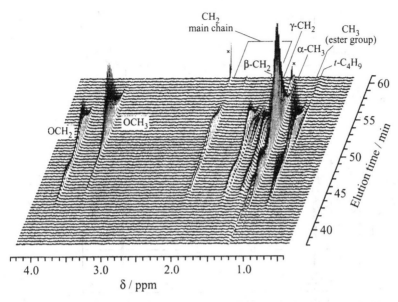

Fig. 8.5. 500 MHz on-line SEC/NMR data of PMMA-*block*-poly(*n*-BuMA) prepared with *t*-C₄H₉MgBr in toluene at −60 °C [8]. The peaks labeled *x* are due to impurities. Eluant CDCl₃, flow rate 0.2 ml/min, amount of sample 1.0 mg, pulse width 60°, pulse repetition time 1.0 s

The mechanism of some stereospecific polymerizations can be elucidated by SEC/NMR [7]. Polymerization of MMA with *t*-C₄H₉Li in toluene at −78 °C gave isotactic-rich PMMA with a broad but unimodal MWD. Addition of trialkylaluminum to *t*-C₄H₉Li changes the initiator stereospecificity for MMA from isotactic-specific to syndiotactic-specific and the polymerization with *t*-C₄H₉Li–(*n*-C₄H₉)₃Al (Al/Li≥2) in toluene at −78 °C gives highly syndiotactic PMMAs [20]. The *t*-C₄H₉Li–(*n*-C₄H₉)₃Al (Al/Li=1) system gives a stereoblock-like PMMA in toluene at −78 °C. The contour plot of the SEC/NMR data for this PMMA is shown in Fig. 8.6. The cross-section at the chemical shift of the OCH₃ peak (Fig. 8.6d) indicates the polymer has a bimodal MWD. The cross-sections at elution times of 49.2 and 57.2 min (Fig. 8.6b, c) clearly show that the lower-molecular-weight part of the polymer is syndiotactic and that the higher-molecular-weight part is isotactic; i.e., isotactic-specific and syndiotactic-specific propagating species exist concomitantly in the polymerization system at a ratio of Al/Li=1.0 [7].

The mechanism of polymerization of MMA by a Grignard reagent was also studied by on-line SEC/NMR. The Grignard reagent exists in the Schlenk equilibrium:

$$2\,RMgX \rightleftharpoons R_2Mg + MgX_2. \tag{8.1}$$

It is reported that $RMgX$ gives highly isotactic PMMA and that R_2Mg gives syndiotactic PMMA with a much higher reactivity than the former [21]. SEC/NMR analysis of the PMMA prepared with *t*-C₄H₉MgBr ([Mg²⁺]/[*t*-C₄H₉Mg-]=0.87)

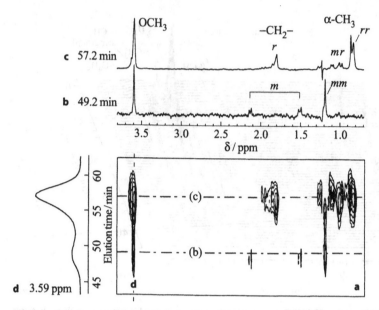

Fig. 8.6. a Contour plot and **b–d** cross-sections of the 500 MHz on-line SEC/NMR data for PMMA prepared with t-C$_4$H$_9$Li–(n-C$_4$H$_9$)$_3$Al (Al/Li=1.0) in toluene at −78 °C [7]. Eluant CDCl$_3$, flow rate 0.2 ml/min, amount of sample 1.0 mg, pulse width 60°, pulse repetition time 1.0 s

in toluene at −78 °C revealed that the polymer had a broad MWD and that the fraction of syndiotactic triad increases with an increase in the molecular weight of the polymer. The results suggest that the higher-molecular-weight part with high syndiotacticity was produced from the species generated by (t-C$_4$H$_9$)$_2$Mg and that the isotactic part, which has a lower molecular weight and a relatively narrow MWD, was produced from the species generated mainly by t-C$_4$H$_9$MgBr [7].

Poly(vinyl alcohol) is usually prepared by the hydrolysis of poly(vinyl acetate) and often contains a small number of unreacted vinyl acetate units. The on-line SEC/NMR analysis of several commercial poly(vinyl alcohol)s in D$_2$O revealed that the number of vinyl acetate units depended on the molecular weight of the polymer; lower-molecular-weight fractions contain larger numbers of acetate units than higher-molecular-weight fractions [13].

8.4 Simple and Accurate Analysis of Oligomers

The on-line SEC/NMR technique also provides structural data for homologous components in oligomers if an SEC column with small porosity is employed. The on-line SEC/NMR data were collected for a mixture of chloral oligomers [9]. The oligomers were prepared by polymerizing chloral with t-C$_4$H$_9$OLi slightly below the ceiling temperature of polymerization and end-capping the anion with an acetate group. The trimer and the higher oligomers have been found to be completely isotactic [22–26] (Structure 8.3).

$$t\text{-}C_4H_9OLi + n\,CCl_3CHO \longrightarrow t\text{-}C_4H_9 \left(\!\!\begin{array}{c} CCl_3 \\ | \\ CH-O \end{array}\!\!\right)_{\!\!n}\!Li$$

Structure 8.3 $\xrightarrow{(CH_3CO)_2O}$ $t\text{-}C_4H_9O\left(\!\!\begin{array}{c} CCl_3 \\ | \\ CH-O \end{array}\!\!\right)_{\!\!n}\!\!\begin{array}{c} O \\ \| \\ CCH_3 \end{array}$

The SEC/NMR-detected chromatogram for the chloral oligomers is shown in Fig. 8.7 together with an RI-detected chromatogram. The ^{1}H NMR detection was carried out by monitoring the peak intensity of the acetal proton signals resonating at 5.92 ppm (trimer), 5.71 ppm (tetramer), 5.84 ppm (pentamer), 5.68 ppm (hexamer), and 6.23 ppm (heptamer). The peaks selected for monitoring were chosen so as to minimize any overlap effect. When the spectra recorded at elution times from 48.6 to 49.2 min are accumulated, for example, the spectrum of the pure pentamer is obtained with a much improved S/N ratio, in spite of the incomplete separation by chromatography. This is one of the great advantages of on-line

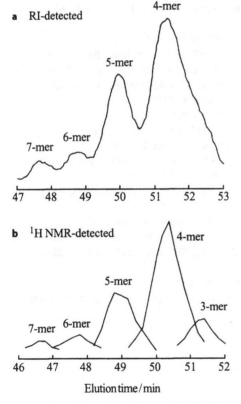

Fig. 8.7. a RI-detected and **b** 500 MHz ^{1}H NMR-detected SEC curves of a mixture of chloral oligomers [9]

SEC/NMR. It should also be noted that in the NMR-detected SEC curves the extent of overlap of the peaks can be easily realized from the chromatogram and the relative peak intensities reflect accurately the molar fractions of the oligomer components. This is another advantage of the on-line SEC/NMR method when applied to the analysis of oligomers [9].

References

1. DORN HC (1984) Anal Chem 56:747A
2. LAUDE DA JR, WILKINS CL (1986) Trends Anal Chem 5:230
3. ALBERT K (1995) J Chromatogr A 703:123
4. ALBERT K (2002) On-line LC-NMR and related techniques. Wiley, New York
5. HATADA K, UTE K, OKAMOTO Y, IMANARI M, FUJII N (1988) Polym Bull 20:317
6. HATADA K, UTE K, KASHIYAMA M, IMANARI M (1990) Polym J 22:218
7. HATADA K, UTE K, KITAYAMA T, NISHIMURA T, KASHIYAMA M, FUJIMOTO N (1990) Polym Bull 23:549
8. HATADA K, UTE K, KITAYAMA T, YAMAMOTO M, NISHIMURA T, KASHIYAMA M (1989) Polym Bull 21:489
9. UTE K, KASHIYAMA M, OKA K, HATADA K, VOGL O (1990) Makromol Chem Rapid Commun 11:31
10. UTE K, NIIMI R, HONGO S, HATADA K (1998) Polym J 30:439
11. UTE K, NIIMI R, HATADA K, KOLBERT AC (1999) Int J Polym Anal Charact 5:47
12. KITAYAMA T, JANCO M, UTE K, NIIMI R, HATADA K, BEREK D (2000) Anal Chem 72:1518
13. HOTTA M, UTE K, HATADA K (1991) Polym Prepr Jpn 40:1123 (English edition E459)
14. HAW JF, GLASS TE, DORN HC (1981) Anal Chem 53:2327
15. ALBERT K, BAYER E (1988) Trends Anal Chem 7:288
16. SMALLCOMBE SH, PATT SL, KEIFER PA (1995) J Magn Reson Ser A 117:295
17. GRUBISIC Z, REMPP P, BENOIT H (1967) J Polym Sci Part B 5:753
18. HATADA K, UTE K, TANAKA K, OKAMOTO Y, KITAYAMA T (1986) Polym J 18:1037
19. KITAYAMA T, TABUCHI M, HATADA K (2000) Polym J 32:796
20. KITAYAMA T, SHINOZAKI T, MASUDA E, YAMAMOTO M, HATADA K (1988) Polym Bull 20:505
21. MATSUZAKI K, TANAKA H, KANAI T (1981) Makromol Chem 182:2905
22. UTE K, NISHIMURA T, HATADA K, XI F, VASS F, VOGL O (1990) Makromol Chem 191:557
23. HATADA K, UTE K, NAKANO T, VASS F, VOGL O (1989) Makromol Chem 190:2217
24. VOGL O (1985) Chemist 62:16
25. ZHANG J, JAYCOX GD, VOGL O (1988) Polymer 29:707
26. VOGL O, UTE K, NISHIMURA T, XI F, VASS F, HATADA K (1989) Macromolecules 22:4658

Subject Index